SAFETY SYMBOLS

SAFETY SYMBOLS	HAZARD	EXAMPLES	PRECAUTION	REMEDY
DISPOSAL	Special disposal procedures need to be followed.	certain chemicals, living organisms	Do not dispose of these materials in the sink or trash can.	Dispose of wastes as directed by your teacher.
BIOLOGICAL	Organisms or other biological materials that might be harmful to humans	bacteria, fungi, blood, unpreserved tissues, plant materials	Avoid skin contact with these materials. Wear mask or gloves.	Notify your teacher if you suspect contact with material. Wash hands thoroughly.
EXTREME TEMPERATURE	Objects that can burn skin by being too cold or too hot	boiling liquids, hot plates, dry ice, liquid nitrogen	Use proper protection when handling.	Go to your teacher for first aid.
SHARP OBJECT	Use of tools or glassware that can easily puncture or slice skin	razor blades, pins, scalpels, pointed tools, dissecting probes, broken glass	Practice common-sense behavior and follow guidelines for use of the tool.	Go to your teacher for first aid.
FUME	Possible danger to respiratory tract from fumes	ammonia, acetone, nail polish remover, heated sulfur, moth balls	Make sure there is good ventilation. Never smell fumes directly. Wear a mask.	Leave foul area and notify your teacher immediately.
ELECTRICAL	Possible danger from electrical shock or burn	improper grounding, liquid spills, short circuits, exposed wires	Double-check setup with teacher. Check condition of wires and apparatus.	Do not attempt to fix electrical problems. Notify your teacher immediately.
IRRITANT	Substances that can irritate the skin or mucous membranes of the respiratory tract	pollen, moth balls, steel wool, fiberglass, potassium permanganate	Wear dust mask and gloves. Practice extra care when handling these materials.	Go to your teacher for first aid.
CHEMICAL	Chemicals that can react with and destroy tissue and other materials	bleaches such as hydrogen peroxide; acids such as sulfuric acid, hydrochloric acid; bases such as ammonia, sodium hydroxide	Wear goggles, gloves, and an apron.	Immediately flush the affected area with water and notify your teacher.
TOXIC	Substance may be poisonous if touched, inhaled, or swallowed	mercury, many metal compounds, iodine, poinsettia plant parts	Follow your teacher's instructions.	Always wash hands thoroughly after use. Go to your teacher for first aid.
OPEN FLAME	Open flame may ignite flammable chemicals, loose clothing, or hair	alcohol, kerosene, potassium permanganate, hair, clothing	Tie back hair. Avoid wearing loose clothing. Avoid open flames when using flammable chemicals. Be aware of locations of fire safety equipment.	Notify your teacher immediately. Use fire safety equipment if applicable.

Eye Safety Proper eye protection should be worn at all times by anyone performing or observing science activities.

Clothing Protection This symbol appears when substances could stain or burn clothing.

Animal Safety This symbol appears when safety of animals and students must be ensured.

Radioactivity This symbol appears when radioactive materials are used.

Glencoe Science

Ecology

NATIONAL GEOGRAPHIC SOCIETY

science.glencoe.com

Glencoe
McGraw-Hill

New York, New York   Columbus, Ohio   Woodland Hills, California   Peoria, Illinois

Glencoe Science

Ecology

Student Edition
Teacher Wraparound Edition
Interactive Teacher Edition CD-ROM
Interactive Lesson Planner CD-ROM
Lesson Plans
Content Outline for Teaching
Dinah Zike's Teaching Science with Foldables
Directed Reading for Content Mastery
Foldables: Reading and Study Skills
Assessment
    Chapter Review
    Chapter Tests
    ExamView Pro Test Bank Software
    Assessment Transparencies
    Performance Assessment in the Science Classroom
    The Princeton Review Standardized Test Practice Booklet
Directed Reading for Content Mastery in Spanish
Spanish Resources
English/Spanish Guided Reading Audio Program
Reinforcement

Enrichment
Activity Worksheets
Section Focus Transparencies
Teaching Transparencies
Laboratory Activities
Science Inquiry Labs
Critical Thinking/Problem Solving
Reading and Writing Skill Activities
Mathematics Skill Activities
Cultural Diversity
Laboratory Management and Safety in the Science Classroom
MindJogger Videoquizzes and Teacher Guide
Interactive CD-ROM with Presentation Builder
Vocabulary PuzzleMaker Software
Cooperative Learning in the Science Classroom
Environmental Issues in the Science Classroom
Home and Community Involvement
Using the Internet in the Science Classroom

THE PRINCETON REVIEW

"Study Tip," "Test-Taking Tip," and the "Test Practice" features in this book were written by The Princeton Review, the nation's leader in test preparation. Through its association with McGraw-Hill, The Princeton Review offers the best way to help students excel on standardized assessments.

The Princeton Review is not affiliated with Princeton University or Educational Testing Service.

Glencoe/McGraw-Hill

A Division of The McGraw·Hill Companies

Cover Images: Three generations of reticulated giraffes are shown in this photo.

Send all inquiries to:
Glencoe/McGraw-Hill
8787 Orion Place
Columbus, OH 43240

ISBN 0-07-825588-0
Printed in the United States of America.
3 4 5 6 7 8 9 10 027/043 06 05 04 03 02

Authors

National Geographic Society
Education Division
Washington, D.C.

Peter Rillero, PhD
Professor of Science Education
Arizona State University West
Phoenix, Arizona

Dinah Zike
Educational Consultant
Dinah-Might Activities, Inc.
San Antonio, Texas

Consultants

Content

Michael A. Hoggarth, PhD
Department of Life and Earth Sciences
Otterbein College
Westerville, Ohio

Dominic Salinas, PhD
Middle School Science Supervisor
Caddo Parish Schools
Shreveport, Louisiana

Safety

Sandra West, PhD
Associate Professor of Biology
Southwest Texas State University
San Marcos, Texas

Reading

Elizabeth Babich
Special Education Teacher
Mashpee Public Schools
Mashpee, Massachusetts

Math

Teri Willard, EdD
Department of Mathematics
Montana State University
Belgrade, Montana

Reviewers

Michelle Bailey
Northwood Middle School
Houston, Texas

Maureen Barrett
Thomas E. Harrington Middle School
Mt. Laurel, New Jersey

Desiree Bishop
Baker High School
Mobile, Alabama

Janice Bowman
Coke R. Stevenson Middle School
San Antonio, Texas

Linda V. Forsyth
Merrill Middle School
Denver, Colorado

Amy Morgan
Berry Middle School
Hoover, Alabama

Michelle Punch
Northwood Middle School
Houston, Texas

Billye Robbins
Jackson Intermediate
Pasadena, Texas

Darcy Vetro-Ravndal
Middleton Middle School of Technology
Tampa, Florida

Series Activity Testers

José Luis Alvarez, PhD
Math/Science Mentor Teacher
El Paso, Texas

Nerma Coats Henderson
Teacher
Pickerington Jr. High School
Pickerington, Ohio

Mary Helen Mariscal-Cholka
Science Teacher
William D. Slider Middle School
El Paso, Texas

José Alberto Marquez
TEKS for Leaders Trainer
El Paso, Texas

Science Kit and Boreal Laboratories
Tonawanda, New York

CONTENTS

Nature of Science: Conservation and Native Americans—2

Interactions of Life—6

SECTION 1 **Living Earth** 8
SECTION 2 **Populations** 12

NATIONAL GEOGRAPHIC   Visualizing Population Growth 18

SECTION 3 **Interactions Within Communities** 20
Activity Feeding Habits of Planaria 25
Activity: Design Your Own Experiment
Population Growth in Fruit Flies 26

TIME *Science and History*
You Can Count On It 28

The Nonliving Environment—34

SECTION 1 **Abiotic Factors** 36
Activity Humus Farm 43
SECTION 2 **Cycles in Nature** 44

NATIONAL GEOGRAPHIC   Visualizing the Carbon Cycle 48

SECTION 3 **Energy Flow** 50
Activity Where does the mass of a plant come from? . . . 54
Science Stats Extreme Climates 56

Ecosystems—62

SECTION 1 **How Ecosystems Change** 64

NATIONAL GEOGRAPHIC   Visualizing Secondary Succession 66

SECTION 2 **Biomes** 68
Activity Studying a Land Ecosystem 76
SECTION 3 **Aquatic Ecosystems** 77
Activity: Use the Internet
Exploring Wetlands 84

TIME *Science and Society*
Helping Nature Help Itself 86

CHAPTER 4

Conserving Resources—92

SECTION 1 **Resources** 94

NATIONAL GEOGRAPHIC Visualizing Solar Energy 101

SECTION 2 **Pollution** 102

Activity The Greenhouse Effect 111

SECTION 3 **The Three Rs of Conservation** 112

Activity: Model and Invent
Solar Cooking 116

Science and Language Arts
Beauty Plagiarized 118

CHAPTER 5

Conserving Life—124

SECTION 1 **Biodiversity** 126

NATIONAL GEOGRAPHIC Visualizing Threatened
and Endangered Species 132

Activity Oily Birds 137

SECTION 2 **Conservation Biology** 138

Activity Biodiversity and the Health of a Plant
Community 144

TIME *Science and Society*
Rain Forest Troubles 146

Field Guides

Biomes Field Guide 156
Waste Management Field Guide 160

Skill Handbooks—164

Reference Handbooks

A. Safety in the Science Classroom 189
B. Care and Use of a Microscope 190
C. Diversity of Life 191

English Glossary—195

Spanish Glossary—199

Index—204

Interdisciplinary Connections/Activities

 NATIONAL GEOGRAPHIC **VISUALIZING**

1 Population Growth 18
2 The Carbon Cycle 48
3 Secondary Succession 66
4 Solar Energy. 101
5 Threatened and Endangered
   Species 132

TIME **SCIENCE AND** *Society*

3 Helping Nature Help Itself 86
5 Rain Forest Troubles 146

TIME **SCIENCE AND HISTORY**

1 You Can Count On It 28

Science Stats

2 Extreme Climates 56

Science **and** Language Arts

4 Beauty Plagiarized. 118

Full Period Labs

1 Feeding Habits of Planaria 25
   Design Your Own Experiment:
   Population Growth in Fruit Flies. . 26
2 Humus Farm 43
   Where does the mass of a
   plant come from? 54
3 Studying a Land Ecosystem 76
   Use the Internet: Exploring
   Wetlands. 84
4 The Greenhouse Effect 111
   Model and Invent: Solar Cooking. . 116
5 Oily Birds 137
   Biodiversity and the Health
   of a Plant Community 144

 **Mini LAB**

1 **Try at Home:** Observing Seedling
   Competition 13
   Comparing Biotic Potential 17
2 **Try at Home:** Determining
   Soil Makeup 38
   Comparing Fertilizers. 47
3 **Try at Home:** Modeling Rain
   Forest Leaves 72
   Modeling Freshwater
   Environments 78
4 **Try at Home:** Observing Mineral
   Mining Effects 96
   Measuring Acid Rain. 103
5 **Try at Home:** Demonstrating
   Divided Habitats 133
   Testing the Effects of Acid Rain . . . 135

Explore Activity

1 Examine sod from a lawn 7
2 Compare climate differences 35
3 Infer the origin of houseplants 63
4 Model topsoil loss 93
5 Recognize environmental
   differences 125

Feature Contents

Problem-Solving Activities

1 Do you have too many crickets?.... 15
4 What items are you recycling
  at home?.................... 114

Math Skills Activities

2 Graphing Temperature
  versus Elevation.............. 40
3 Calculating Temperatures........ 80
5 Analyzing Data for Biodiversity... 129

Skill Builder Activities

Science
Classifying: 53
Communicating: 11, 49, 83, 115, 143
Comparing and Contrasting: 110
Concept Mapping: 67, 83, 100
Forming Hypotheses: 11
Identifying and Manipulating Variables
  and Controls: 24, 42, 49, 136
Making and Using Tables: 19, 115
Recognizing Cause and Effect: 143
Recording Observations: 75

Math
Solving One-Step Equations: 19, 53, 67,
  100, 110

Technology
Using a Database: 75, 136
Using an Electronic Spreadsheet: 42
Using Graphics Software: 24

Science
INTEGRATION

Chemistry: 21, 79
Earth Science: 51, 74, 140
Health: 106
Physics: 41, 97

SCIENCE *Online*

Research: 10, 49, 81, 104, 114, 142
Data Update: 16, 41, 65, 139

THE
PRINCETON
REVIEW

Test Practice: 33, 61, 91, 123, 151, 152–153

Conservation and Native Americans

Figure 1
Salmon were the major food source of the Pacific Northwest Native Americans.

Figure 2
Native Americans did not waste any part of nature. This Blackfoot shirt was made with hair, porcupine quills, and feathers.

A cool breeze blows through the trees lining the Columbia River in the state of Washington as a shiny salmon darts through the water. Every year these fish make the difficult journey upstream to lay their eggs in the same waters where they began their lives.

Many years ago, before Europeans came to North America, Native Americans of the northwest depended on salmon. At the first sign of autumn, the salmon run would start. Native American oral histories tell of rivers being so full of the fish that you almost could walk across them to the other side. The native people harvested and dried enough fish to sustain them for the entire year. They also had a deep respect for the salmon. The belief that the fish had a spirit was shown by the tradition of thanking the salmon's spirit for its sacrifice before it was eaten.

Respect for Life

Native American tradition is to respect all animals and their spirits that gave their life to feed and clothe the people. On the Great Plains, the Cheyenne and other native peoples had strict rules against killing more bison than they needed. They believed in using every part of the animal. They used as much of the animal as possible for food. The animal's fat was used in cooking. Tools were made from bones. Clothing, shoes, and blankets were made from hides. Even the bisons' stomachs were used as water pouches.

Respect for life extended to plants and crops for Native Americans. The Iroquois of the northeastern United States celebrated festivals in honor of the "three sisters"—corn, squash, and beans—their essential foods. The Maya who once lived in what is now southern Mexico and Central America, felt that if someone cut down a tree unnecessarily, that person's life would be shortened.

Every year, in the late spring or early summer, native peoples living on the Plains, including the Cree, Kiowa, Shoshone, and others, celebrated the cycle of life with the Sun Dance. This ceremony, which is performed by many tribes today, stresses the regeneration of life and humans' connection to Earth. Native Americans expressed thankfulness for Earth's gifts—the return of flowers and crops in the spring and summer, and the sacrifice made by animal spirits as they give their bodies to sustain the people. Their tradition recognizes that people must cooperate with nature so that revival and rebirth can continue.

Figure 3
Agriculture students experiment with the Native American practice of growing several types of plants in one area.

Respect for Earth

Native Americans felt a deep connection to Earth. While some European settlers viewed the New World as a wilderness ready to be tamed, the native peoples believed in living in harmony with Earth. They didn't understand the European desire to own and develop land. To the Native Americans, everyone shared the land. You could no more own the land than you could own the air. The Lakota chief Black Elk once spoke of each human's responsibility to Earth in this way: "Every step that we take upon You should be done in a sacred manner; every step should be taken as a prayer."

Figure 4
Native American artist Helen Hardin titled this piece *Father Sky Embracing Mother Earth.*

Figure 5
Early industrial development led to environmental pollution.

Need for Conservation

When Europeans began to settle throughout America, it seemed that their way of life would completely replace Native American conservation practices. Many farming techniques used by the new settlers eroded fertile topsoil. Some of the wildlife, which had been treasured and respected by the native peoples, was driven to either extinction or near extinction.

New technology greatly affected the environment. The steam and diesel engines, commercial oil drilling, burning of coal, and industrial development led to waste and pollution.

In recent years, Americans have become more aware of the need for conservation. In 1970, the Environmental Protection Agency was created with the goal of safeguarding the environment. Paper, plastic, glass, and metal are widely recycled. Many items that were once considered waste are now being reused. Although pollution is still a problem, Americans are returning to practices that are more like traditional Native American ways.

Science

The practice of modifying human behavior to preserve Earth is known as conservation. Conservation involves managing the use of natural resources found in many different environments. In this book, you will learn about how all species on Earth are connected and how they can affect their environments.

History of Science

History is a record of people's achievements and their mistakes. Science history teaches that scientific ideas are not limited to scientists. Many people of different ethnic backgrounds, professions, and ages—male and female—have contributed to today's conservation movement. Studying the history of science reminds people that even though accepted views may change, humans constantly gain a better understanding of nature.

Figure 6
Recycling and using nonpolluting types of energy, such as solar energy, are two of the ways Americans are becoming more environmentally conscious.

Connections

Throughout their history, Native Americans have understood the importance of living in harmony with nature. They practiced conservation as a way of life, not just a passing fad. The people of the northwest understood that fishing for more salmon than they needed would endanger the next year's supply. Today, people are learning this same lesson as some traditional fishing grounds are closed because of overharvesting.

Recall the Mayan belief about how unnecessarily cutting down a tree shortened one's life. It is now known that plants provide oxygen and remove carbon dioxide from the air. Scientists warn that cutting down rain forest trees might lead to a buildup of carbon dioxide in the atmosphere. This could add to the trend in rising, global average temperatures known as global warming.

Native Americans on the Plains realized the importance of using every part of the bison, learning to live with their minimum needs, and limiting their use of natural resources. This model is being followed today, in some ways, through the effort to reduce, reuse, and recycle.

As new knowledge replaces old, it's tempting to think that only new ideas are worthwhile. Native American traditions of conservation show that this is not always true. These traditions once were disregarded but the value of their conservation is now known.

Figure 7
Many items that you use every day are made of recycled material. These packing beads are made from a wheat product.

Figure 8
The amount of land covered by rain forests is decreasing daily.

You Do It

Forests originally covered about 25 percent of Earth's land areas. Today, only about thirteen percent of Earth's land areas are covered with forests. Research to find out the history of this change. Explain how following Native American conservation practices might have avoided the problems that face forests today.

Interactions of Life

Why would a powerful rhinoceros allow birds to perch on its back? Why aren't these birds safely perched in a tree? How do they find food? You don't have to go to Africa to see birds on the back of a rhino. You can see these animals at zoos or wildlife parks. In this chapter, you will learn how living organisms interact with each other and their surroundings. You also will learn about the roles each organism plays in the flow of energy through the environment.

What do you think?

Science Journal Look at the picture below with a classmate. Discuss what you think this might be or what is happening. Here's a hint: *It's a city within a city.* Write your answer or best guess in your Science Journal.

EXPLORE ACTIVITY

In your lifetime, you probably have taken thousands of footsteps on grassy lawns or playing fields. If you take a close look at the grass, you'll see that each blade is attached to roots in the soil. How do the grass plants obtain everything they need to live and grow? What other kinds of organisms live in the grass? The following activity will give you a chance to take a closer look at the life in a lawn.

Examine sod from a lawn

1. Examine a section of sod from a lawn.
2. How do the roots of the grass plants hold the soil?
3. Do you see signs of other living things besides grass?

Observe

In your Science Journal, answer the above questions and describe any organisms that are present in your section of sod. Explain how these organisms might affect the growth of grass plants. Draw a picture of your section of sod.

Before You Read

FOLDABLES
Reading & Study Skills

Making a Concept Map Study Fold The following Foldable will help you organize information by diagramming ideas about your favorite wild animal.

1. Place a sheet of paper in front of you with the short side at the top. Fold the paper in half from the left side to the right side.
2. Fold from top to bottom to divide the paper into thirds, then open up the three folds.
3. Through the top thickness of paper, cut along each of the fold lines to the side fold, forming three tabs.
4. Label *Organism, Population,* and *Community* across the front of the paper, as shown. Write the name of your favorite wild animal under the *Organism* tab.
5. Before you read the chapter, write what you know about your favorite animal under the top tab. As you read the chapter, write how this animal is part of a population and a community under the middle and bottom tabs.

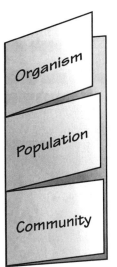

SECTION 1

Living Earth

As You Read

What You'll Learn

- **Identify** places where life is found on Earth.
- **Define** ecology.
- **Observe** how the environment influences life.

Vocabulary

biosphere
ecosystem
ecology

population
community
habitat

Why It's Important

All living things on Earth depend on each other for survival.

The Biosphere

What makes Earth different from other planets in the solar system? One difference is Earth's abundance of living organisms. The part of Earth that supports life is the **biosphere** (BI uh sfihr). The biosphere includes the top portion of Earth's crust, all the waters that cover Earth's surface, and the atmosphere that surrounds Earth.

✔ **Reading Check** *What three things make up the biosphere?*

As **Figure 1** shows, the biosphere is made up of different environments that are home to different kinds of organisms. For example, desert environments receive little rain. Cactus plants, coyotes, and lizards are included in the life of the desert. Tropical rain forest environments receive plenty of rain and warm weather. Parrots, monkeys, and tens of thousands of other organisms live in the rain forest. Coral reefs form in warm, shallow ocean waters. Arctic regions near the north pole are covered with ice and snow. Polar bears, seals, and walruses live in the arctic.

Figure 1
Earth's biosphere consists of many environments, including ocean waters, polar regions, and deserts.

Desert

Arctic

Coral reef

Life on Earth In our solar system, Earth is the third planet from the Sun. The amount of energy that reaches Earth from the Sun helps make the temperature just right for life. Mercury, the planet closest to the Sun, is too hot during the day and too cold at night to make life possible there. Venus, the second planet from the Sun, has a thick, carbon dioxide atmosphere and high temperatures. It is unlikely that life could survive there. Mars, the fourth planet, is much colder than Earth because it is farther from the Sun and has a thinner atmosphere. It might support microscopic life, but none has been found. The planets beyond Mars probably do not receive enough heat and light from the Sun to have the right conditions for life.

Ecosystems

On a visit to Yellowstone National Park in Wyoming, you might see a prairie scene like the one shown in **Figure 2.** Bison graze on prairie grass. Cowbirds follow the bison, catching grasshoppers that jump away from the bisons' hooves. This scene is part of an ecosystem. An **ecosystem** consists of all the organisms living in an area and the nonliving features of their environment. Bison, grass, birds, and insects are living organisms of this prairie ecosystem. Water, temperature, sunlight, soil, and air are nonliving features of this prairie ecosystem. **Ecology** is the study of interactions that occur among organisms and their environment. Ecologists are scientists who study these interactions.

✔ **Reading Check** *What is an ecosystem?*

Figure 2
Ecosystems are made up of living organisms and the nonliving features of their environment. In this prairie ecosystem, cowbirds eat insects and bison graze on grass. *What other kinds of organisms might live in this ecosystem?*

Populations

Suppose you meet an ecologist who studies how a herd of bison moves from place to place and how the female bison in the herd care for their young. This ecologist is studying the members of a population. A **population** is made up of all the organisms in an ecosystem that belong to the same species. For example, all the bison in a prairie ecosystem are one population. All the cowbirds in this ecosystem make up a different population. The grasshoppers make up yet another population.

Ecologists often study how populations interact. For example, an ecologist might try to answer questions about several prairie species. How does grazing by bison affect the growth of prairie grass? How does grazing influence the insects that live in the grass and the birds that eat those insects? This ecologist is studying a community. A **community** refers to all the populations in an ecosystem. The prairie community is made of populations of bison, grasshoppers, cowbirds, and all other species in the prairie ecosystem. An arctic community might include populations of fish, seals that eat fish, and polar bears that hunt and eat seals. **Figure 3** shows how organisms, populations, communities, and ecosystems are related.

SCIENCE Online

Research Visit the Glencoe Science Web site at **science.glencoe.com** and find out the estimated human population size for the world today. In your Science Journal, create a graph that shows the population change between the year 2000 and this year.

Figure 3
The living world is arranged in several levels of organization.

Figure 4
The trees of the forest provide a habitat for woodpeckers and other birds. This salamander's habitat is the moist forest floor.

Habitats

Each organism in an ecosystem needs a place to live. The place in which an organism lives is called its **habitat.** The animals shown in **Figure 4** live in a forest ecosystem. Trees are the woodpecker's habitat. These birds use their strong beaks to pry insects from tree bark or break open acorns and nuts. Woodpeckers usually nest in holes in dead trees. The salamander's habitat is the forest floor, beneath fallen leaves and twigs. Salamanders avoid sunlight and seek damp, dark places. This animal eats small worms, insects, and slugs. An organism's habitat provides the kinds of food and shelter, the temperature, and the amount of moisture the organism needs to survive.

Section 1 Assessment

1. What is the biosphere?
2. What is ecology?
3. How are the terms *habitat* and *biosphere* related to each other?
4. What is the major difference between a community and a population? Give one example of each.
5. **Think Critically** Does the amount of rain that falls in an area determine which kinds of organisms can live there? Why or why not?

Skill Builder Activities

6. **Forming Hypotheses** Make a hypothesis about how one nonliving feature of an ecosystem would affect the growth of dandelions in that ecosystem. **For more help, refer to the Science Skill Handbook.**

7. **Communicating** Pretend you are a non-human organism in the wild. Describe what you are and list living and nonliving features of the environment that affect you. **For more help, refer to the Science Skill Handbook.**

Populations

As You Read

What You'll Learn

- **Identify** methods for estimating population sizes.
- **Explain** how competition limits population growth.
- **List** factors that influence changes in population size.

Vocabulary
limiting factor
carrying capacity

Why It's Important
Competition caused by population growth affects many organisms, including humans.

Competition

Some pet shops sell lizards, snakes, and other reptiles. Crickets are raised as a food supply for pet reptiles. In the wild, crickets come out at night and feed on plant material. During the day, they hide in dark areas, beneath leaves or under buildings. Pet shop workers who raise crickets make sure that the insects have plenty of food, water, and hiding places. As the cricket population grows, the workers increase the crickets' food supply and the number of hiding places. To avoid crowding, some of the crickets could be moved into larger containers.

Food and Space Organisms living in the wild do not always have enough food or living space. The Gila woodpecker, shown in **Figure 5,** lives in the Sonoran Desert of Arizona and Mexico. This bird makes its nest in a hole that it drills in a saguaro (suh GWAR oh) cactus. If an area has too many Gila woodpeckers or too few saguaros, the woodpeckers must compete with each other for nesting spots. Competition occurs when two or more organisms seek the same resource at the same time.

Growth Limits Competition limits population size. If the amount of available nesting space is limited, some woodpeckers will not be able to raise young. Gila woodpeckers eat cactus fruit, berries, and insects. If food becomes scarce, some woodpeckers might not survive to reproduce. Competition for food, living space, or other resources can prevent population growth.

In nature, the most intense competition is usually among individuals of the same species, because they need the same kinds of food and shelter. Competition also takes place among individuals of different species. For example, after a Gila woodpecker has abandoned its nesting hole, owls, flycatchers, snakes, and lizards compete for the shelter of the empty hole.

Figure 5
Gila woodpeckers make nesting holes in the saguaro cactus. Many animals compete for the shelter these holes provide.

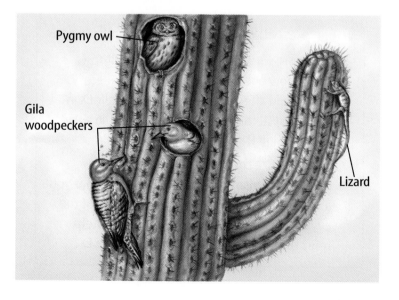

Pygmy owl

Gila woodpeckers

Lizard

Population Size

Ecologists often need to measure the size of a population. This information can indicate whether or not a population is healthy and growing. Population counts can help identify populations that could be in danger of disappearing.

Some populations are easy to measure. If you were raising crickets, you could measure the size of your cricket population simply by counting all the crickets in the container. What if you wanted to compare the cricket populations in two different containers? You would calculate the number of crickets per square meter (m^2) of your container. The size of a population that occupies a specific area is called population density. **Figure 6** shows human population density in different places in the world.

✔ **Reading Check** *What is population density?*

Measuring Populations Counting crickets can be tricky. They look alike, move a lot, and hide. The same cricket could be counted more than once, and others could be completely missed. Ecologists have similar problems when measuring wildlife populations. One of the methods they use is called trap-mark-release. Suppose you want to count wild rabbits. Rabbits live underground and come out at dawn and dusk to eat. Ecologists set traps that capture rabbits without injuring them. Each captured rabbit is marked and released. Later, another sample of rabbits is captured. Some of these rabbits will have marks, but many will not. By comparing the number of marked and unmarked rabbits in the second sample, ecologists can estimate the population size.

TRY AT HOME
Mini LAB

Observing Seedling Competition

Procedure
1. Fill **two plant pots** with **moist potting soil.**
2. Plant **radish seeds** in one pot, following the spacing instructions on the seed packet. Label this pot "Recommended Spacing."
3. Plant radish seeds in the second pot, spaced half the recommended distance apart. Label this pot "Densely Populated." Wash your hands.
4. Keep the soil moist. When the seeds sprout, move them to a well-lit area.
5. Measure the height of the seedlings every two days for two weeks. Record the data in your **Science Journal.**

Analysis
1. Which plants grew faster?
2. Which plants looked healthiest after two weeks?
3. How did competition influence the plants?

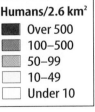

Humans/2.6 km²	
■	Over 500
■	100–500
■	50–99
□	10–49
□	Under 10

Figure 6
This map shows human population density. *Which countries have the highest population density?*

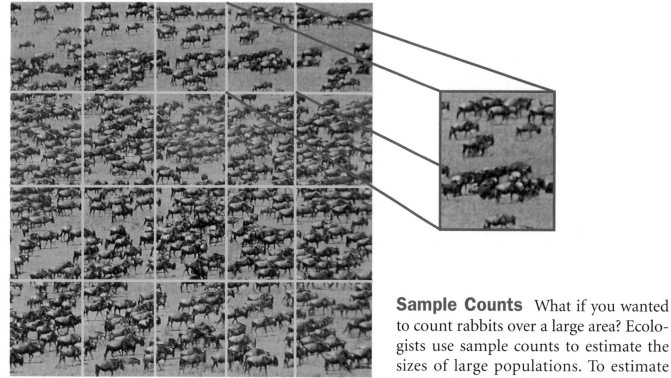

Figure 7
Ecologists can estimate population size by making a sample count. Wildebeests graze on the grassy plains of Africa. *How could you use the enlarged square to estimate the number of wildebeests in the entire photograph?*

Sample Counts What if you wanted to count rabbits over a large area? Ecologists use sample counts to estimate the sizes of large populations. To estimate the number of rabbits in a 100-acre area, for example, you could count the rabbits in one acre and multiply by 100 to estimate the population size. **Figure 7** shows another approach to sample counting.

Limiting Factors One grass plant can produce hundreds of seeds. Imagine those seeds drifting onto a vacant field. Many of the seeds sprout and grow into grass plants that produce hundreds more seeds. Soon the field is covered with grass. Can this grass population keep growing forever? Suppose the seeds of wildflowers or trees drift onto the field. If those seeds sprout, trees and flowers would compete with grasses for sunlight, soil, and water. Even if the grasses did not have to compete with other plants, they might eventually use up all the space in the field. When no more living space is available, the population cannot grow.

In any ecosystem, the availability of food, water, living space, mates, nesting sites, and other resources is often limited. A **limiting factor** is anything that restricts the number of individuals in a population. Limiting factors include living and nonliving features of the ecosystem.

A limiting factor can affect more than one population in a community. Suppose a lack of rain limits plant growth in a meadow. Fewer plants produce fewer seeds. For seed-eating mice, this reduction in the food supply could become a limiting factor. A smaller mouse population could, in turn, become a limiting factor for the hawks and owls that feed on mice.

Carrying Capacity A population of robins lives in a grove of trees in a park. Over several years, the number of robins increases and nesting space becomes scarce. Nesting space is a limiting factor that prevents the robin population from getting any larger. This ecosystem has reached its carrying capacity for robins. **Carrying capacity** is the largest number of individuals of one species that an ecosystem can support over time. If a population begins to exceed the environment's carrying capacity, some individuals will not have enough resources. They could die or be forced to move elsewhere, like the deer shown in **Figure 8.**

Figure 8
These deer might have moved into a residential area because a nearby forest's carrying capacity for deer has been reached.

✔ **Reading Check** *How are limiting factors related to carrying capacity?*

Problem-Solving Activity

Do you have too many crickets?

You've decided to raise crickets to sell to pet stores. A friend says you should not allow the cricket population density to go over 210 crickets/m^2. Use what you've learned in this section to measure the population density in your cricket tanks.

Identifying the Problem

The table on the right lists the areas and populations of your three cricket tanks. How can you determine if too many crickets are in one tank? If a tank contains too many crickets, what could you do? Explain why too many crickets in a tank might be a problem.

Cricket Population		
Tank	Area (m^2)	Number of Crickets
1	0.80	200
2	0.80	150
3	1.5	315

Solving the Problem

1. Do any of the tanks contain too many crickets? Could you make the population density of the three tanks equal by moving crickets from one tank to another? If so, which tank would you move crickets into?

2. The population density of wild crickets living in a field is 2.4 crickets/m^2. If the field has an area of 250 m^2, what is the approximate size of the cricket population? Why would the population density of crickets in a field be lower than the population density of crickets in a tank?

Biotic Potential What would happen if no limiting factors restricted the growth of a population? Think about a population that has an unlimited supply of food, water, and living space. The climate is favorable. Population growth is not limited by diseases, predators, or competition with other species. Under ideal conditions like these, the population would continue to grow.

The highest rate of reproduction under ideal conditions is a population's biotic potential. The larger the number of offspring that are produced by parent organisms, the higher the biotic potential of the species will be. Compare an avocado tree to a tangerine tree. Assume that each tree produces the same number of fruits. Each avocado fruit contains one large seed. Each tangerine fruit contains a dozen seeds or more. Because the tangerine tree produces more seeds per fruit, it has a higher biotic potential than the avocado tree.

Changes in Populations

Birthrates and death rates also influence the size of a population and its rate of growth. A population gets larger when the number of individuals born is greater than the number of individuals that die. When the number of deaths is greater than the number of births, populations get smaller. Take the squirrels living in New York City's Central Park as an example. In one year, if 900 squirrels are born and 800 die, the population increases by 100. If 400 squirrels are born and 500 die, the population decreases by 100.

The same is true for human populations. **Table 1** shows birthrates, death rates, and population changes for several countries around the world. In countries with faster population growth, birthrates are much higher than death rates. In countries with slower population growth, birthrates are only slightly higher than death rates. In Germany, where the population is getting smaller, the birthrate is lower than the death rate.

Table 1 Population Growth			
	Birthrate*	**Death Rate***	**Population Increase** (percent)
Rapid-Growth Countries			
Jordan	38.8	5.5	3.3
Uganda	50.8	21.8	2.9
Zimbabwe	34.3	9.4	5.2
Slow-Growth Countries			
Germany	9.4	10.8	−1.5
Sweden	10.8	10.6	0.1
United States	14.8	8.8	0.6

*Number per 1,000 people

Figure 9
The mangrove seeds sprout while they are still attached to the parent tree. Some sprouted seeds drop into the mud below the parent tree and continue to grow. Others drop into the water and can be carried away by tides and ocean currents. When they wash ashore, they might start a new population of mangroves or add to an existing mangrove population.

Moving Around Most animals can move easily from place to place, and these movements can affect population size. For example, a male mountain sheep might wander many miles in search of a mate. After he finds a mate, their offspring might establish a completely new population of mountain sheep far from the male's original population.

Many bird species move from one place to another during their annual migrations. During the summer, populations of Baltimore orioles are found throughout eastern North America. During the winter, these populations disappear because the birds migrate to Central America. They spend the winter there, where the climate is mild and food supplies are plentiful. When summer approaches, the orioles migrate back to North America.

Even plants and microscopic organisms can move from place to place, carried by wind, water, or animals. The tiny spores of mushrooms, mosses, and ferns float through the air. The seeds of dandelions, maple trees, and other plants have feathery or winglike growths that allow them to be carried by wind. Spine-covered seeds hitch rides by clinging to animal fur or people's clothing. Many kinds of seeds can be transported by river and ocean currents. Mangrove trees growing along Florida's Gulf Coast, shown in **Figure 9,** provide an example of how water moves seeds.

Mini LAB

Comparing Biotic Potential

Procedure
1. Remove all the seeds from a **whole fruit.** Do not put fruit or seeds in your mouth.
2. Count the total number of seeds in the fruit. Wash your hands, then record these data in your Science Journal.
3. Compare your seed totals with those of classmates who examined other types of fruit.

Analysis
1. Which type of fruit had the most seeds? Which had the fewest seeds?
2. What is an advantage of producing many seeds? Can you think of a possible disadvantage?
3. To estimate the total number of seeds produced by a tomato plant, what would you need to know?

Figure 10

When a species enters an ecosystem that has abundant food, water, and other resources, its population can flourish. Beginning with a few organisms, the population increases until the number of organisms and available resources are in balance. At that point, population growth slows or stops. A graph of these changes over time produces an S-curve, as shown here for coyotes.

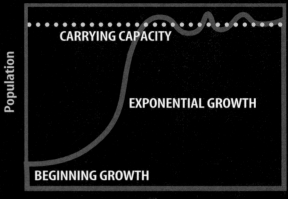

CARRYING CAPACITY

EXPONENTIAL GROWTH

BEGINNING GROWTH

Population

Time

BEGINNING GROWTH During the first few years, population growth is slow, because there are few adults to produce young. As the population grows, so does the number of breeding adults.

EXPONENTIAL GROWTH As the number of adults in the population grows, so does the number of births. The coyote population undergoes exponential growth, quickly increasing in size.

CARRYING CAPACITY As resources become less plentiful, the birthrate declines and the death rate may rise. Population growth slows. The coyote population has reached the environmental carrying capacity—the maximum number of coyotes that the environment can sustain.

Exponential Growth

Imagine what might happen if a pair of coyotes moves into a valley where no other coyotes live. Food and water are abundant, and there are plenty of areas where female coyotes can build dens for their young. This population grows quickly in a pattern called exponential growth. Exponential growth means that the larger a population becomes, the faster it grows.

After several years, the population becomes so large that the coyotes begin to compete for food and den sites. Population growth slows, and the number of coyotes remains fairly constant and reaches equilibrium. This ecosystem has reached its carrying capacity for coyotes. A graph that describes each stage in this pattern of population growth is shown in **Figure 10.** As you can see in **Figure 11,** Earth's human population shows exponential growth. In the year 2000, Earth's human population exceeded 6 billion. By the year 2050, it is estimated that Earth's human population could reach 10 billion.

Increase in Human Population

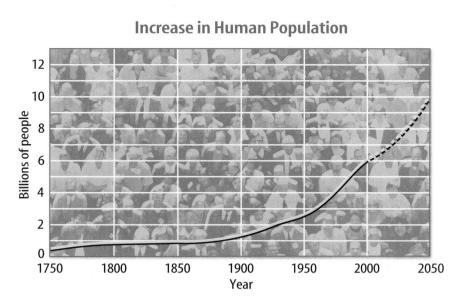

Figure 11
The size of the human population is increasing by about 1.6 percent per year. *What factors affect human population growth?*

Section ② Assessment

1. How can an ecologist predict the size of a population without counting every organism in the population?

2. Why does competition between individuals of the same species tend to be greater than competition between individuals of different species?

3. How do birthrates and death rates influence the size of a population?

4. How does carrying capacity influence the number of organisms in an ecosystem?

5. **Think Critically** Why does the supply of food and water in an ecosystem usually affect population size more than other limiting factors?

Skill Builder Activities

6. **Making and Using Tables** Construct a table using the following data on changes in the size of a deer population in Arizona. In 1910 there were 6 deer; in 1915, 36 deer; in 1920, 143 deer; in 1925, 86 deer; and in 1935, 26 deer. Propose a hypothesis to explain what might have caused these changes. **For more help, refer to the** Science Skill Handbook.

7. **Solving One-Step Equations** A vacant lot that measures 12 m × 12 m contains 46 dandelion plants, 212 grass plants, and 14 bindweed plants. What is the population density, per square meter, of each species? **For more help, refer to the** Math Skill Handbook.

3 Interactions Within Communities

Obtaining Energy

Just as a car engine needs a constant supply of gasoline, living organisms need a constant supply of energy. The energy that fuels most life on Earth comes from the Sun. Some organisms use the Sun's energy to create energy-rich molecules through the process of photosynthesis. The energy-rich molecules, usually sugars, serve as food. They are made up of different combinations of carbon, hydrogen, and oxygen atoms. Energy is stored in the chemical bonds that hold the atoms of these molecules together. When the molecules break apart—for example, during digestion—the energy in the chemical bonds is released to fuel life processes.

Producers Organisms that use an outside energy source like the Sun to make energy-rich molecules are called **producers.** Most producers contain chlorophyll (KLOR uh fihl), a chemical that is required for photosynthesis. As shown in **Figure 12,** green plants are producers. Some producers do not contain chlorophyll and do not use energy from the Sun. Instead, they make energy-rich molecules through a process called chemosynthesis (kee moh SIHN thuh sus). These organisms can be found near volcanic vents on the ocean floor. Inorganic molecules in the water provide the energy source for chemosynthesis.

Figure 12
Green plants, including the grasses that surround this pond, are producers. The pond also contains many other producers, including microscopic organisms like **A** *Euglena* and **B** simple plantlike organisms called algae.

A Magnification: 125×

B Magnification: 225×

Consumers

Herbivores

Carnivores

Omnivores

Decomposers

Figure 13
Four categories of consumers are shown. *What kind of consumer is a cactus wren? A mushroom?*

Consumers Organisms that cannot make their own energy-rich molecules are called **consumers.** Consumers obtain energy by eating other organisms. **Figure 13** shows the four general categories of consumers. Herbivores are the vegetarians of the world. They include rabbits, deer, and other plant eaters. Carnivores are animals that eat other animals. Frogs and spiders are carnivores that eat insects. Omnivores, including pigs and humans, eat mostly plants and animals. Decomposers, including fungi, bacteria, and earthworms, consume wastes and dead organisms. Decomposers help recycle once-living matter by breaking it down into simple, energy-rich substances. These substances might serve as food for decomposers, be absorbed by plant roots, or be consumed by other organisms.

✔ **Reading Check** *How are producers different from consumers?*

Food Chains Ecology includes the study of how organisms depend on each other for food. A food chain is a simple model of the feeding relationships in an ecosystem. For example, shrubs are food for deer, and deer are food for mountain lions, as illustrated in **Figure 14.** What food chain would include you?

Chemistry
INTEGRATION

Glucose is a nutrient molecule produced during photosynthesis. Look up the chemical structure of glucose and draw it in your Science Journal.

Figure 14
Food chains illustrate how consumers obtain energy from other organisms in an ecosystem.

Symbiotic Relationships

Not all relationships among organisms involve food. Many organisms live together and share resources in other ways. Any close relationship between species is called **symbiosis**.

Figure 15
Many examples of symbiotic relationships exist in nature.

A Lichens are a result of mutualism.

Mutualism You may have noticed crusty lichens growing on fences, trees, or rocks. Lichens, like those shown in **Figure 15A,** are made up of an alga or a cyanobacterium that lives within the tissues of a fungus. Through photosynthesis, the cyanobacterium or alga supplies energy to itself and the fungus. The fungus provides a protected space in which the cyanobacterium or alga can live. Both organisms benefit from this association. A symbiotic relationship in which both species benefit is called **mutualism** (MYEW chuh wuh lih zum).

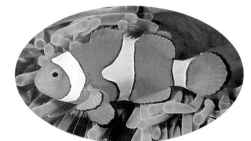

B Clown fish and sea anemones have a commensal relationship.

Commensalism If you've ever visited a marine aquarium, you might have seen the ocean organisms shown in **Figure 15B.** The creature with gently waving, tubelike tentacles is a sea anemone. The tentacles contain a mild poison. Anemones use their tentacles to capture shrimp, fish, and other small animals to eat. The striped clown fish can swim among the tentacles without being harmed. The anemone's tentacles protect the clown fish from predators. In this relationship, the clown fish benefits but the sea anemone is not helped or hurt. A symbiotic relationship in which one organism benefits and the other is not affected is called **commensalism** (kuh MEN suh lih zum).

Magnification: 128×

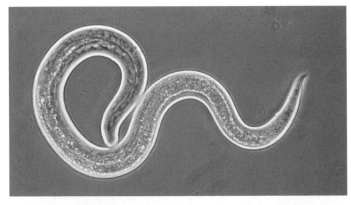

C Some roundworms are parasites that rob nutrients from their hosts.

Parasitism Pet cats or dogs sometimes have to be treated for worms. Roundworms, like the one shown in **Figure 15C,** are common in puppies. This roundworm attaches itself to the inside of the puppy's intestine and feeds on nutrients in the puppy's blood. The puppy may have abdominal pain, bloating, and diarrhea. If the infection is severe, the puppy might die. A symbiotic relationship in which one organism benefits but the other is harmed is called **parasitism** (PER uh suh tih zum).

Niches

One habitat might contain hundreds or even thousands of species. Look at the rotting log habitat shown in **Figure 16.** A rotting log in a forest can be home to many species of insects, including termites that eat decaying wood and ants that feed on the termites. Other species that live on or under the rotting log include millipedes, centipedes, spiders, and worms. You might think that competition for resources would make it impossible for so many species to live in the same habitat. However, each species has different requirements for its survival. As a result, each species has its own niche (NIHCH). A **niche** refers to how an organism survives, how it obtains food and shelter, how it finds a mate and cares for its young, and how it avoids danger.

✔ **Reading Check** *Why does each species have its own niche?*

Special adaptations that improve survival are often part of an organism's niche. Milkweed plants contain a poison that prevents many insects from feeding on them. Monarch butterfly caterpillars have an adaptation that allows them to eat milkweed. Monarchs can take advantage of a food resource that other species cannot use. Milkweed poison also helps protect monarchs from predators. When the caterpillars eat milkweed, they become slightly poisonous. Birds avoid eating monarchs because they learn that the caterpillars and adult butterflies have an awful taste and can make them sick.

Health

INTEGRATION

The poison in milkweed is similar to the drug digitalis. Small amounts of digitalis are used to treat heart ailments in humans, but it is poisonous in large doses. Look up digitalis and explain in your Science Journal how it affects the human body.

Figure 16
Different adaptations enable each species living in this rotting log to have its own niche.
🅐 Termites eat wood. They make tunnels inside the log.
🅑 Millipedes feed on plant matter and find shelter beneath the log. 🅒 Wolf spiders capture insects living in and around the log.

Figure 17
The alligator is a predator. The turtle is its prey.

Predator and Prey When you think of survival in the wild, you might imagine an antelope running away from a lion. An organism's niche includes how it avoids being eaten and how it finds or captures its food. Predators, like the one shown in **Figure 17,** are consumers that capture and eat other consumers. The prey is the organism that is captured by the predator. The presence of predators usually increases the number of different species that can live in an ecosystem. Predators limit the size of prey populations. As a result, food and other resources are less likely to become scarce, and competition between species is reduced.

Cooperation Individual organisms often cooperate in ways that improve survival. For example, a white-tailed deer that detects the presence of wolves or coyotes will alert the other deer in the herd. Many insects, such as ants and honeybees, live in social groups. Different individuals perform different tasks required for the survival of the entire nest. Soldier ants protect workers that go out of the nest to gather food. Worker ants feed and care for ant larvae that hatch from eggs laid by the queen. These cooperative actions improve survival and are a part of the species' niche.

Section 3 Assessment

1. Explain why all consumers ultimately depend on producers for food.
2. Draw a food chain that models the feeding relationships of three species in a community. Choose a food chain other than the one shown in **Figure 14.**
3. Make up two imaginary organisms that have a mutualistic relationship. Give them names and explain how they benefit from the association.
4. What is the difference between a habitat and a niche?
5. **Think Critically** A parasite can obtain food only from a host organism. Most parasites weaken but do not kill their hosts. Why?

Skill Builder Activities

6. **Manipulating Variables and Controls** You are sure that Animal A benefits from a relationship with Plant B, but you are not sure if Plant B benefits, is harmed, or is unaffected by the relationship. Design an experiment to compare how well Plant B grows on its own and when Animal A is present. **For more help, refer to the Science Skill Handbook.**
7. **Using Graphics Software** Use graphics software to make three different food chains. Represent each organism with a shape that resembles it. For example, you could use a leaf shape to represent a plant. Label each shape. **For more help,** refer to the Technology Skill Handbook.

Activity

Feeding Habits of Planaria

You probably have watched minnows darting about in a stream. It is not as easy to observe organisms that live at the bottom of a stream, beneath rocks, logs, and dead leaves. Countless stream organisms, including insect larvae, worms, and microscopic organisms, live out of your view. One such organism is a type of flatworm called a planarian. In this activity, you will find out about the eating habits of planarians.

What You'll Investigate
What food items do planarians prefer to eat?

Materials
small bowl
planarians (several)
lettuce leaf
raw liver or meat
guppies (several)
pond or stream water
magnifying lens

Goals
■ **Observe** the food preference of planarians.
■ **Infer** what planarians eat in the wild.

Safety Precautions

Procedure
1. Fill the bowl with stream water.
2. Place a lettuce leaf, piece of raw liver, and several guppies in the bowl. Add the planarians. Wash your hands.
3. **Observe** what happens inside the bowl for at least 20 minutes. Do not disturb the bowl or its contents. Use a magnifying lens to look at the planarians.
4. **Record** all of your observations in your Science Journal.

Conclude and Apply
1. Which food did the planarians prefer?
2. **Infer** what planarians might eat when in their natural environment.
3. Based on your observations during this activity, what is a planarian's niche in a stream ecosystem?
4. **Predict** where in a stream you might find planarians. Use references to find out whether your prediction is correct.

Communicating Your Data

Share your results with other students in your class. Plan an adult-supervised trip with several classmates to a local stream to search for planarians in their native habitat. **For more help, refer to the** Science Skill Handbook.

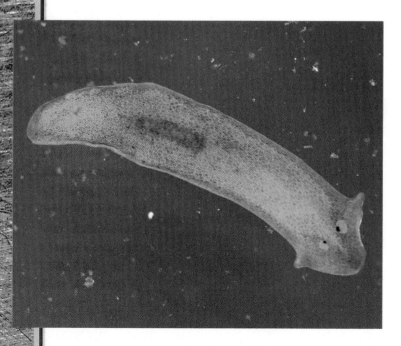

Population Growth in Fruit Flies

Populations can grow at an exponential rate only if the environment provides the right amount of food, shelter, air, moisture, heat, living space, and other factors. You probably have seen fruit flies hovering near ripe bananas or other fruit. Fruit flies are fast-growing organisms often raised in science laboratories. The flies are kept in culture tubes and fed a diet of specially prepared food flakes. Can you improve on this standard growing method to achieve faster population growth?

Recognize the Problem

Will a change in one environmental factor affect the growth of a fruit fly population?

Form a Hypothesis

Based on your reading about fruit flies, state a hypothesis about how changing one environmental factor will affect the rate of growth of a fruit fly population.

Goals
- **Identify** the environmental factors needed by a population of fruit flies.
- **Design** an experiment to investigate how a change in one environmental factor affects in any way the size of a fruit fly population.
- **Observe** and **measure** changes in population size.

Possible Materials
fruit flies
standard fruit fly culture kit
food items (banana, orange peel, or other fruit)
water
heating or cooling source
culture containers
cloth, plastic, or other tops for culture containers
hand lens

Safety Precautions

Test Your Hypothesis

Plan

1. As a group, decide on one environmental factor to investigate. Agree on a hypothesis about how a change in this factor will affect population growth. Decide how you will test your hypothesis, and identify the experimental results that would support your hypothesis.

2. **List** the steps you will need to take to test your hypothesis. Describe exactly what you will do. List your materials.

3. **Determine** the method you will use to measure changes in the size of your fruit fly populations.

4. Prepare a data table in your Science Journal to record weekly measurements of your fruit fly populations.

5. Read the entire experiment and make sure all of the steps are in a logical order.

6. **Research** the standard method used to raise fruit flies in the laboratory. Use this method as the control in your experiment.

7. **Identify** all constants, variables, and controls in your experiment.

Do

1. Make sure your teacher approves your plan before you start.

2. Carry out your experiment.

3. **Measure** the growth of your fruit fly populations weekly and record the data in your data table.

Analyze Your Data

1. What were the constants in your experiment? The variables?

2. **Compare** changes in the size of your control population with changes in your experimental population. Which population grew faster?

3. Using the information in your data table, make a line graph that shows how the sizes of your two fruit fly populations changed over time. Use a different colored pencil for each population's line on the graph.

Draw Conclusions

1. Did the results support your hypothesis? Explain.

2. **Compare** the growth of your control and experimental populations. Did either population reach exponential growth? How do you know?

Communicating Your Data

Compare the results of your experiment with those of other students in your class. **For more help, refer to the** Science Skill Handbook.

YOU CAN COUNT

The Census gives a snapshot of the people of the United States

The doorbell rings and you hear someone at the door say to your mom, "I'm working for the U.S. Census Bureau, doing follow-up interviews. Do you have a few minutes to answer some questions?" What does this person—and the U.S. government—want to know about your family?

Counting people is important to the United States and to many other countries around the world. It helps governments determine the distribution of people in the various regions of a nation. To obtain this information, the government takes a census—a count of how many people are living in their country on a particular day at a particular time, and in a particular place. A census is a snapshot of a country's population. The time at which the count occurs is called the "census moment." Some countries close their borders for a day or two so everyone will "sit still" for the census camera at the census moment, as was done in Nigeria in 1991.

Counting on the Count

When the United States government was formed, its founders set up the House of Representatives based on population. Areas with more people had more government representatives, and areas with fewer people had fewer representatives. In 1787, the requirement for a census became part of the Constitution. A census must be taken every ten years so the proper number of representatives for each state can be calculated.

Over the years, the U.S. Census Bureau has added questions to obtain more information than just a population count. In 1810, questions about manufacturing were added. In 1850, as more immigrants began coming to the United States, a new question about where people were born was added. In 1880, census takers asked people whether or not they were married. And in 1950, the first electronic computers were used to add up the census results.

Next, read on to find out more about the census.

ON IT

Growing by the Numbers

Chances are you just blinked your eyes. While you did it, three people were added to the world's population. There, you blinked again—that's another three people! It may seem impossible, but that's how quickly the world's population is growing. It adds up to 184 people every minute, 11,040 every hour, 264,960 every day, and 97 million every year! On October 12, 1999, the official number of people on the planet reached a record 6 billion.

The Short Form

Before 1970, United States census data was collected by field workers. They went door to door to count the number of people living in each household. Since then, the census has been done mostly by mail. People are sent a form they must fill out. The form asks for the number of people living at an address and their names, races, ages, and relationships. Answers to these and other questions are confidential. Census workers visit some homes to check on the accuracy of the information. The census helps the government to figure out how the population is aging. Census data are also important in deciding how to distribute government services and funding.

The 2000 Snapshot

One of the findings of the 2000 Census is that the U.S. population is becoming more equally spread out across age groups. By analyzing the data from the census, officials estimate that by 2020 the population of children, middle-aged people, and senior citizens will be about equal. It's predicted also that there will be more people who are over 100 years old than ever before.

Martha F. Riche researches population changes in the United States. She was also a director of the Census Bureau. Riche thinks that the more equal distribution in age will lead to challenges for the nation. How will we meet the demands of more people who are living longer? Will we need to build more hospitals to care for them? Will more children mean a need to build more schools? Federal, state, and local governments will be using the results of the 2000 Census for years to come as they plan our future.

Martha F. Riche studies population changes.

CONNECTIONS Census Develop a school census. What questions will you ask? (Don't ask questions that are too personal.) Who will ask them? How will you make sure you counted everyone? Using the results, can you make any predictions about your school's future or its current students?

SCIENCE

Online

For more information, visit science.glencoe.com

Reviewing Main Ideas

Section 1 Living Earth

1. Ecology is the study of interactions that take place in the biosphere. *Is ice-covered Antarctica a part of Earth's biosphere? Why or why not?*

2. Populations are made up of all organisms of the same species living in an area.

3. Communities are made up of all the populations of different species of organisms living in one ecosystem.

4. Living and nonliving factors affect an organism's ability to survive in its habitat.

Section 2 Populations

1. Population size can be estimated by counting a sample of a total population.

2. Competition for limiting factors can restrict the size of a population. *What limiting factors might influence the size of a rabbit population?*

3. Population growth is affected by birthrate, death rate, and the movement of individuals into or out of a community.

4. Exponential population growth can occur in environments that provide a species with plenty of food, shelter, and other resources.

Section 3 Interactions Within Communities

1. All life requires energy.

2. Most producers use the Sun's energy to make food in the form of energy-rich molecules. Consumers obtain their food by eating other organisms.

3. Mutualism, commensalism, and parasitism are the three kinds of symbiosis.

4. Every species has its own niche, which includes adaptations for survival. *What adaptations are involved in the relationship between the milkweed plant and the caterpillar of the monarch butterfly?*

FOLDABLES
Reading & Study Skills

After You Read

Under the population tab of your Concept Map Study Fold, write what would happen if there were an increase in the population of your animal.

Visualizing Main Ideas

Complete the following concept map on communities.

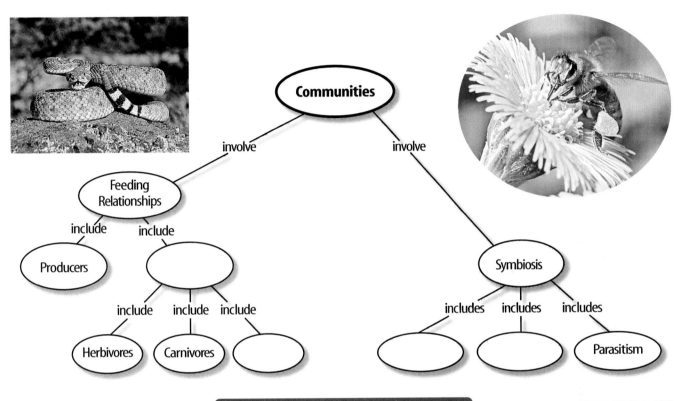

Vocabulary Review

Vocabulary Words

- **a.** biosphere
- **b.** carrying capacity
- **c.** commensalism
- **d.** community
- **e.** consumer
- **f.** ecology
- **g.** ecosystem
- **h.** habitat
- **i.** limiting factor
- **j.** mutualism
- **k.** niche
- **l.** parasitism
- **m.** population
- **n.** producer
- **o.** symbiosis

THE PRINCETON REVIEW **Study Tip**

Get together with a friend to study. Quiz each other about specific topics from your textbook and class material to prepare for a test.

Using Vocabulary

Explain the difference between the vocabulary words in each of the following sets.

1. niche, habitat
2. mutualism, commensalism
3. limiting factor, carrying capacity
4. biosphere, ecosystem
5. producer, consumer
6. population, ecosystem
7. community, population
8. parasitism, symbiosis
9. ecosystem, ecology
10. parasitism, commensalism

Chapter ① Assessment

Checking Concepts

Choose the word or phrase that best answers the question.

1. Which of the following is a living factor in the environment?
   A) animals   C) sunlight
   B) air   D) soil

2. What is made up of all the populations in an area?
   A) niches   C) community
   B) habitats   D) ecosystem

3. What does the number of individuals in a population that occupies an area of a specific size describe?
   A) clumping   C) spacing
   B) size   D) density

4. Which of the following animals is an example of an herbivore?
   A) wolf   C) tree
   B) moss   D) rabbit

5. What term best describes a symbiotic relationship in which one species is helped and the other is harmed?
   A) mutualism   C) commensalism
   B) parasitism   D) consumerism

6. Which of the following conditions tends to increase the size of a population?
   A) births exceed deaths
   B) population size exceeds the carrying capacity
   C) movements out of an area exceed movements into the area
   D) severe drought

7. Which of the following is most likely to be a limiting factor in a population of fish living in the shallow water of a large lake?
   A) sunlight   C) food
   B) water   D) soil

8. An ecologist wants to know the size of a population of wild daisy plants growing in a meadow. The meadow measures 1,000 m². The ecologist counts 30 daisy plants in a sample area that is 100 m². What is the estimated population of daisies in the entire meadow?
   A) 3   C) 300
   B) 30   D) 3,000

9. Which of these organisms is a producer?
   A) mole   C) whale
   B) owl   D) oak tree

10. Which pair of words is incorrect?
    A) black bear—carnivore
    B) grasshopper—herbivore
    C) pig—omnivore
    D) lion—carnivore

Thinking Critically

11. Why does a parasite have a harmful effect on the organism it infects?

12. What factors affect carrying capacity?

13. Describe your own habitat and niche.

14. The female cowbird lays eggs in the nest of another bird. The other birds care for and feed the cowbird chicks when they hatch. Which type of symbiosis is this?

15. Explain how several different niches can exist in the same habitat.

Developing Skills

16. **Making Models** Place the following organisms in the correct sequence to model a food chain: grass, snake, mouse, and hawk.

17. Predicting Dandelion seeds can float great distances on the wind with the help of white, featherlike attachments. Predict how a dandelion seed's ability to be carried on the wind helps reduce competition among dandelion plants.

18. Classifying Classify the following relationships as parasitism, commensalism, or mutualism: a shark and a remora fish that cleans and eats parasites from the shark's gills; head lice and a human; a spiny sea urchin and a tiny fish that hides from predators by floating among the sea urchin's spines.

19. Comparing and Contrasting Compare and contrast the diets of omnivores and herbivores. Give examples of each.

20. Making and Using Tables Complete the following table.

Types of Symbiosis

Organism A	Organism B	Relationship
Gains	Doesn't gain or lose	
Gains		Mutualism
Gains	Loses	

Performance Assessment

21. Poster Use photographs from old magazines to create a poster that shows at least three different food chains. Display your poster for your classmates.

TECHNOLOGY

Go to the Glencoe Science Web site at **science.glencoe.com** or use the **Glencoe Science CD-ROM** for additional chapter assessment.

THE PRINCETON REVIEW **Test Practice**

A food web shows how organisms in a particular ecosystem depend on each other for food. The food web below shows how the plants and animals in a grassland ecosystem obtain energy from each other.

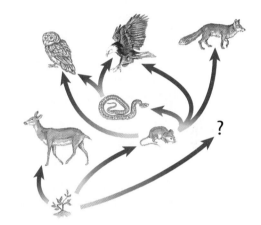

Study the picture and answer the following questions.

1. Other organisms also live in this habitat. Which of the following organisms could fill in the blank space in this food web?
   A) tree
   B) bison
   C) alligator
   D) hawk

2. Suppose all the snakes were removed from this ecosystem. Which of the following statements represents the most reasonable prediction of what could happen in this ecosystem?
   F) The plants would die.
   G) The owls would start eating foxes.
   H) There would be no more predators to eat the mice.
   J) The eagles would start eating more mice.

The Nonliving Environment

Could you write a story about what would happen if the Sun stopped shining? Most life on Earth depends on the Sun's energy. In this chapter, you'll learn about how organisms called producers use energy to make food and how other organisms called consumers take in that food. You'll also read about cycles in nature such as the water, carbon, and nitrogen cycles, and many other nonliving factors that affect your life.

What do you think?

Science Journal Look at the picture below with a classmate. Discuss what this might be. Here's a hint: *It's a factory that relies on sunlight for its energy supply.* Write your answer or best guess in your Science Journal.

Do you live in a dry, sandy region covered with cactus plants or desert scrub? Is your home in the mountains? Does snow fall during the winter? Perhaps you live near the coast, where flowers bloom year-round. Earth has many ecosystems. In this chapter, you'll learn why the nonliving factors in each ecosystem are different. The following activity will get you started.

Compare climate differences

1. Locate your city or town on a globe or world map. Find your latitude. Latitude shows your distance from the equator and is expressed in degrees, minutes, and seconds.

2. Locate another city with the same latitude as your city but on a different continent.

3. Locate a third city with latitude close to the equator.

4. Using references, compare average annual precipitation and average high and low temperatures for all three cities.

Observe

In your Science Journal, hypothesize how latitude affects average temperatures and rainfall.

Before You Read

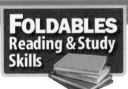

FOLDABLES
Reading & Study Skills

Making a Cause and Effect Study Fold Make the following Foldable to help you understand the cause and effect relationship of the nonliving environment.

Nonliving
Water
Soil
Wind
Temperature
Elevation

1. Place a sheet of paper in front of you so the long side is at the top. Fold the left and right sides in to divide the paper into thirds. Then fold it in half from left to right. Unfold all the folds.

2. Using the fold lines as a guide, refold the paper into a fan. Unfold all the folds again.

3. Before you read the chapter, draw a picture of a familiar ecosystem on one side of the paper. On the other side, label the folds *Nonliving, Water, Soil, Wind, Temperature,* and *Elevation* as shown.

4. As you read the chapter, write on the folds how each nonliving factor affects the environment you drew.

Abiotic Factors

As You Read

What You'll Learn

- **Identify** common abiotic factors in most ecosystems.
- **List** the components of air that are needed for life.
- **Explain** how climate influences life in an ecosystem.

Vocabulary

biotic   soil
abiotic   climate
atmosphere

Why It's Important

Knowing how organisms depend on the nonliving world can help humans maintain a healthy environment.

Environmental Factors

Living organisms depend on one another for food and shelter. The leaves of plants provide food and a home for grasshoppers, caterpillars, and other insects. Many birds depend on insects for food. Dead plants and animals decay and become part of the soil. The features of the environment that are alive, or were once alive, are called **biotic** (bi AH tihk) factors. The term *biotic* means "living."

Biotic factors are not the only things in an environment that are important to life. Most plants cannot grow without sunlight, air, water, and soil. Animals cannot survive without air, water, or the warmth that sunlight provides. The nonliving, physical features of the environment are called **abiotic** (ay bi AH tihk) factors. The prefix *a* means "not." The term *abiotic* means "not living." Abiotic factors include air, water, soil, sunlight, temperature, and climate. The abiotic factors in an environment often determine which kinds of organisms can live there. For example, water is an important abiotic factor in the environment, as shown in **Figure 1.**

Figure 1
Abiotic factors—air, water, soil, sunlight, temperature, and climate—influence all life on Earth.

Air

Air is invisible and plentiful, so it is easily overlooked as an abiotic factor of the environment. The air that surrounds Earth is called the **atmosphere.** Air contains 78 percent nitrogen, 21 percent oxygen, 0.94 percent argon, 0.03 percent carbon dioxide, and trace amounts of other gases. Some of these gases provide substances that support life.

Carbon dioxide (CO_2) is required for photosynthesis. Photosynthesis—a series of chemical reactions—uses CO_2, water, and energy from sunlight to produce sugar molecules. Organisms like plants that can use photosynthesis are called producers because they produce their own food. During photosynthesis, oxygen is released into the atmosphere.

When a candle burns, oxygen from the air chemically combines with the molecules of candle wax. Chemical energy stored in the wax is converted and released as heat and light energy. In a similar way, cells use oxygen to release the chemical energy stored in sugar molecules. This process is called respiration. Through respiration, cells obtain the energy needed for all life processes. Air-breathing animals aren't the only organisms that need oxygen. Plants, some bacteria, algae, fish, and most other organisms also need oxygen for respiration.

Water

Water is essential to life on Earth. It is a major ingredient of the fluid inside the cells of all organisms. In fact, most organisms are 50 percent to 95 percent water. Respiration, digestion, photosynthesis, and many other important life processes can take place only in the presence of water. As **Figure 2** shows, environments that have plenty of water usually support a greater diversity of and a larger number of organisms than environments that have little water.

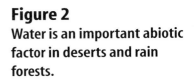

Figure 2
Water is an important abiotic factor in deserts and rain forests.

A Life in deserts is limited to species that can survive for long periods without water.

B Thousands of species can live in lush rain forests where rain falls almost every day.

Determining Soil Makeup

Procedure

1. Collect 2 cups of **soil.** Remove large pieces of debris and break up clods.
2. Put the soil in a **quart jar or similar container that has a lid.**
3. Fill the container with **water** and add 1 teaspoon of **dishwashing liquid.**
4. Put the lid on tightly and shake the container.
5. After 1 min, measure and record the depth of sand that settled on the bottom.
6. After 2 h, measure and record the depth of silt that settles on top of the sand.
7. After 24 h, measure and record the depth of the layer between the silt and the floating organic matter.

Analysis

1. Clay particles are so small that they can remain suspended in water. Where is the clay in your sample?
2. Is sand, silt, or clay the greatest part of your soil sample?

Soil

Soil is a mixture of mineral and rock particles, the remains of dead organisms, water, and air. It is the topmost layer of Earth's crust, and it supports plant growth. Soil is formed, in part, of rock that has been broken down into tiny particles.

Soil is considered an abiotic factor because most of it is made up of nonliving rock and mineral particles. However, soil also contains living organisms and the decaying remains of dead organisms. Soil life includes bacteria, fungi, insects, and worms. The decaying matter found in soil is called humus. Soils contain different combinations of sand, clay, and humus. The type of soil present in a region has an important influence on the kinds of plant life that grow there.

Sunlight

All life requires energy, and sunlight is the energy source for almost all life on Earth. During photosynthesis, producers convert light energy into chemical energy that is stored in sugar molecules. Consumers are organisms that cannot make their own food. Energy is passed to consumers when they eat producers or other consumers. As shown in **Figure 3,** photosynthesis cannot take place if light is never available.

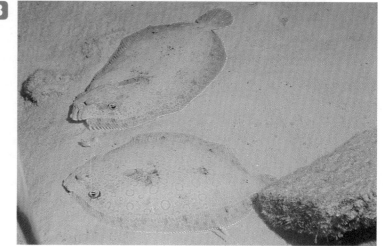

Figure 3

Photosynthesis requires light. **A** Little sunlight reaches the shady forest floor, so plant growth beneath trees is limited. **B** Sunlight does not reach into deep lake or ocean waters. Photosynthesis can take place only in shallow water or near the water's surface. *How do fish that live at the bottom of the deep ocean obtain energy?*

Figure 4
Temperature is an abiotic factor that can affect an organism's survival.

 **A** The penguin has a thick layer of fat to hold in heat and keep the bird from freezing. These emperor penguins huddle together for added warmth.

B The Arabian camel stores fat only in its hump. This way, the camel loses heat from other parts of its body, which helps it stay cool in the hot desert.

Temperature

Sunlight supplies life on Earth with light energy for photosynthesis and heat energy for warmth. Most organisms can survive only if their body temperatures stay within the range of 0°C to 50°C. Water freezes at 0°C. The penguins in **Figure 4** are adapted for survival in the freezing Antarctic. Camels can survive the hot temperatures of the Arabian Desert because their bodies are adapted for staying cool. The temperature of a region depends in part on the amount of sunlight it receives. The amount of sunlight depends on the land's latitude and elevation.

Figure 5
Because Earth is curved, latitudes farther from the equator are colder than latitudes near the equator.

✔ **Reading Check** *What does sunlight provide for life on Earth?*

Latitude In this chapter's Explore Activity, you discovered that temperature is affected by latitude. You found that cities located at latitudes farther from the equator tend to have colder temperatures than cities at latitudes nearer to the equator. As **Figure 5** shows, polar regions receive less of the Sun's energy than equatorial regions. Near the equator, sunlight strikes Earth directly. Near the poles, sunlight strikes Earth at an angle, which spreads the energy over a larger area.

Figure 6
The stunted growth of these trees is a result of abiotic factors.

Elevation If you have climbed or driven up a mountain, you probably noticed that the temperature got cooler as you went higher. A region's elevation, or distance above sea level, affects its temperature. Earth's atmosphere acts as insulation that traps the Sun's heat. At higher elevations, the atmosphere is thinner than it is at lower elevations. Air becomes warmer when sunlight heats the air molecules. Because there are fewer air molecules at higher elevations, air temperatures there tend to be cooler.

Figure 6 shows how elevation affects other abiotic conditions, including soil and wind. At higher elevations, trees are shorter and the ground is rocky. Above the timberline—the elevation beyond which trees do not grow—plant life is limited to low-growing plants. The tops of some mountains are so cold that no plants can survive. Some mountain peaks are covered with snow year-round.

Math Skills Activity

Graphing Temperature Versus Elevation

Example Problem

You climb a mountain and record the temperature every 1,000 m of elevation. The temperature is 30°C at 304.8 m, 25°C at 609.6 m, 20°C at 914.4 m, 15°C at 1,219.2 m, and 5°C at 1,828.8 m. Make a graph of the data. Use your graph to predict the temperature at an altitude of 2,133.6 m.

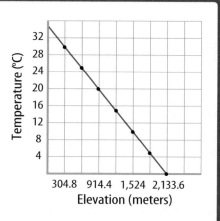

Solution

1. *This is what you know:*
   The data can be written as ordered pairs (elevation, temperature). The ordered pairs for these data are (304.8, 30), (609.6, 25), (914.4, 20), (1,219.2, 15), (1,828.8, 5).

2. *This is what you want to find:*
   Predict the temperature at an elevation of 2,133.6 m.

3. *This is what you need to do:*
   Graph the data by plotting elevation on the x-axis and temperature on the y-axis. Draw a line to connect the data points on your graph.

4. *Predict the temperature at 2,133.6 m:*
   Extend the graph line to predict the temperature at 2,133.6 m.

   Practice Problem

   Temperatures on another mountain are 33°C at sea level, 31°C at 125 m, 29°C at 250 m, and 26°C at 425 m. Graph the data and predict the temperature at 550 m.

For more help, refer to the Math Skill Handbook.

Climate

In Fairbanks, Alaska, winter temperatures may be as low as −52°C, and more than a meter of snow might fall in one month. In Key West, Florida, snow never falls and winter temperatures rarely dip below 5°C. These two cities have different climates. **Climate** refers to an area's average weather conditions over time, including temperature, rainfall or other precipitation, and wind.

For the majority of living things, temperature and precipitation are the two most important components of climate. The average temperature and rainfall in an area influence the type of life found there. Suppose a region has an average temperature of 25°C and receives an average of less than 25 cm of rain every year. It is likely to be the home of cactus plants and other desert life. A region with similar temperatures that receives more than 300 cm of rain every year is probably a tropical rain forest.

Wind Heat energy from the Sun not only determines temperature, but also is responsible for the wind. The air is made up of molecules of gas. As the temperature increases, the molecules spread farther apart. As a result, warm air is lighter than cold air. Colder air sinks below warmer air and pushes it upward, as shown in **Figure 7.** These motions create air currents that are called wind.

Physics
INTEGRATION

Gravity pulls the gases of the atmosphere toward Earth's surface. Also, the weight of the air at the top of the atmosphere presses down on the air below it. In your Science Journal, explain why air at sea level is thicker than air at the top of a mountain.

SCIENCE *Online*

Data Update Visit the Glencoe Science Web site at **science.glencoe.com** to look up recent weather data for your area. In your Science Journal, describe how these weather conditions affect plants or animals that live in your area.

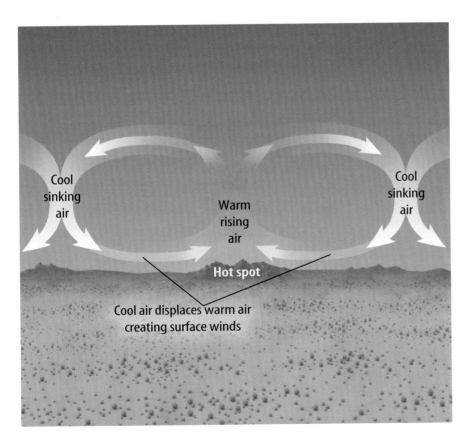

Cool
sinking
air

Warm
rising
air

Cool
sinking
air

Hot spot

Cool air displaces warm air
creating surface winds

Figure 7
Winds are created when sunlight heats some portions of Earth's surface more than others. In areas that receive more heat, the air becomes warmer. Cold air sinks beneath the warm air, forcing the warm air upward.

Figure 8
In Washington State, the western side of the Cascade Mountains receives an average of 101 cm of rain each year. The eastern side of the Cascades is in a rain shadow that receives only about 25 cm of rain per year.

Earth Science
INTEGRATION

The Rain Shadow Effect The presence of mountains can affect rainfall patterns. As **Figure 8** shows, wind blowing toward one side of a mountain is forced upward by the mountain's shape. As the air nears the top of the mountain, it cools. When air cools, the moisture it contains falls as rain or snow. By the time the cool air crosses over the top of the mountain, it has lost most of its moisture. The other side of the mountain range receives much less precipitation. It is not uncommon to find lush forests on one side of a mountain range and desert on the other side.

Section 1 Assessment

1. What is the difference between biotic and abiotic factors?
2. What substances in the air are required for life on Earth?
3. Why is soil considered an abiotic factor and a biotic factor?
4. Why is climate an important abiotic factor?
5. **Think Critically** On day 1 of a hiking trip, you walk in shade under tall trees. On day 2, the trees are shorter and farther apart. On day 3, you see small plants but no trees. On day 4, you see snow. What abiotic factors might contribute to these changes?

Skill Builder Activities

6. **Identifying and Manipulating Variables and Controls** Describe an experiment to find out how much water different types of dry soil can hold. **For more help, refer to the** Science Skill Handbook.

7. **Using an Electronic Spreadsheet** Obtain two months of temperature and precipitation data for two cities in your state. Enter the data in a spreadsheet and calculate average daily temperature and rainfall. Use your calculations to compare the two climates. **For more help, refer to the** Technology Skill Handbook.

Activity

Humus Farm

Soil contains abiotic factors, including rock particles and minerals. Soil also contains biotic factors, such as bacteria, molds, fungi, worms, insects, and decayed organisms. The crumbly, dark brown soil found in gardens or forests contains a high percentage of humus. Humus is formed primarily from the decayed remains of plants, animals, and animal droppings. It adds essential nutrients to the soil, including nitrogen. In this activity, you will cultivate your own humus.

What You'll Investigate
How does humus form?

Materials
widemouth jar	water
soil	marker
grass clippings or green leaves	metric ruler
	graduated cylinder

Goals
- **Observe** the formation of humus.
- **Observe** biotic factors in the soil.
- **Infer** how humus forms naturally.

Safety Precautions 🧤 🥽 🚫
Wash your hands thoroughly after handling soil, grass clippings, or leaves.

Humus Formation	
Date	**Observations**

Procedure
1. Copy the data table below into your Science Journal.
2. Place 4 cm of soil in the jar. Pour 30 mL of water into the jar to moisten the soil.
3. Place 2 cm of grass clippings or green leaves on top of the soil in the jar.
4. Use a marker to mark the height of the grass clippings or green leaves in the jar.
5. Put the jar in a sunny place. Every other day, add 30 mL of water to it. In your Science Journal, write a prediction of what you think will happen in your jar.
6. **Observe** your jar every other day for four weeks. Record your observations in your data table.

Conclude and Apply
1. **Describe** what happened during your investigation.
2. **Infer** how molds and bacteria help the process of humus formation.
3. **Infer** how humus forms on forest floors or in grasslands.

𝒞ommunicating Your Data

Compare your humus farm with those of your classmates. With several classmates, write a recipe for creating the richest humus. Ask your teacher to post your recipe in the classroom. **For more help, refer to the Science Skill Handbook.**

2 Cycles in Nature

As You Read

What You'll Learn

- **Explain** the importance of Earth's water cycle.
- **Diagram** the carbon cycle.
- **Recognize** the role of nitrogen in life on Earth.

Vocabulary

evaporation   nitrogen fixation
condensation   nitrogen cycle
water cycle   carbon cycle

Why It's Important

The recycling of matter on Earth demonstrates natural processes.

Figure 9
Water vapor is a gas that is present in the atmosphere.

A **Water evaporates after a summer rain.**

The Cycles of Matter

Imagine an aquarium tank containing water, fish, snails, plants, algae, and bacteria. The tank is sealed so that only light can enter. Food, water, and air cannot be added. Will the organisms in this environment survive? Through photosynthesis, plants and algae produce their own food. They also supply oxygen to the tank. Fish and snails take in oxygen and eat plants and algae. Wastes from fish and snails fertilize plants and algae. Organisms that die are decomposed by the bacteria. The organisms in this closed environment can survive because the materials are recycled. A constant supply of light energy is the only requirement. Earth's biosphere also contains a fixed amount of water, carbon, nitrogen, oxygen, and other materials required for life. These materials cycle through the environment and are reused by different organisms.

Water Cycle

If you leave a glass of water on a sunny windowsill, the water will disappear. It evaporates. **Evaporation** takes place when liquid water changes into water vapor, which is a gas, and enters the atmosphere, as shown in **Figure 9.** Water evaporates from the surfaces of lakes, streams, puddles, and oceans. Water vapor enters the atmosphere from plant leaves in a process known as transpiration (trans puh RAY shun). Animals release water vapor into the air when they exhale. Water also returns to the environment from animal wastes.

B **Water also evaporates from the ocean.**

Transpiration

Precipitation

Condensation

Evaporation

Groundwater

Condensation Water vapor that has been released into the atmosphere eventually comes into contact with colder air. The temperature of the water vapor drops. Over time, the water vapor cools enough to change back into liquid water. The process of changing from a gas to a liquid is called **condensation.** Water vapor condenses on particles of dust in the air, forming tiny droplets. At first, the droplets clump together to form clouds. When they become large and heavy enough, they fall to the ground as rain or other precipitation. As the diagram in **Figure 10** shows, the **water cycle** is a model that describes how water moves from the surface of Earth to the atmosphere and back to the surface again.

Water Use **Table 1** gives data on the amount of water people take from reservoirs, rivers, and lakes for use in households, businesses, agriculture, and power production. These actions can reduce the amount of water that evaporates into the atmosphere. They also can influence how much water returns to the atmosphere by limiting the amount of water available to plants and animals.

Figure 10
The water cycle involves evaporation, condensation, and precipitation. Water molecules can follow several pathways through the water cycle. *How many water cycle pathways can you identify from this diagram?*

Table 1 U.S. Estimated Water Use in 1990

Water Use	Millions of Gallons per Day	Percent of Total
Homes and Businesses	39,100	11.5
Industry and Mining	27,800	8.2
Farms and Ranches	141,000	41.5
Electricity Production	131,800	38.6

Nitrogen Cycle

The element nitrogen is important to all living things. Nitrogen is a necessary ingredient of proteins. Proteins are required for the life processes that take place in the cells of all organisms. Nitrogen is also an essential part of the DNA of all organisms. Although nitrogen is the most plentiful gas in the atmosphere, most organisms cannot use nitrogen directly from the air. Plants need nitrogen that has been combined with other elements to form nitrogen compounds. Through a process called **nitrogen fixation,** some types of soil bacteria can form the nitrogen compounds that plants need. Plants absorb these nitrogen compounds through their roots. Animals obtain the nitrogen they need by eating plants or other animals. When dead organisms decay, the nitrogen in their bodies returns to the soil or to the atmosphere. This transfer of nitrogen from the atmosphere to the soil, to living organisms, and back to the atmosphere is called the **nitrogen cycle,** shown in **Figure 11.**

✔ Reading Check *What is nitrogen fixation?*

Figure 11
During the nitrogen cycle, nitrogen gas from the atmosphere is converted to a soil compound that plants can use.

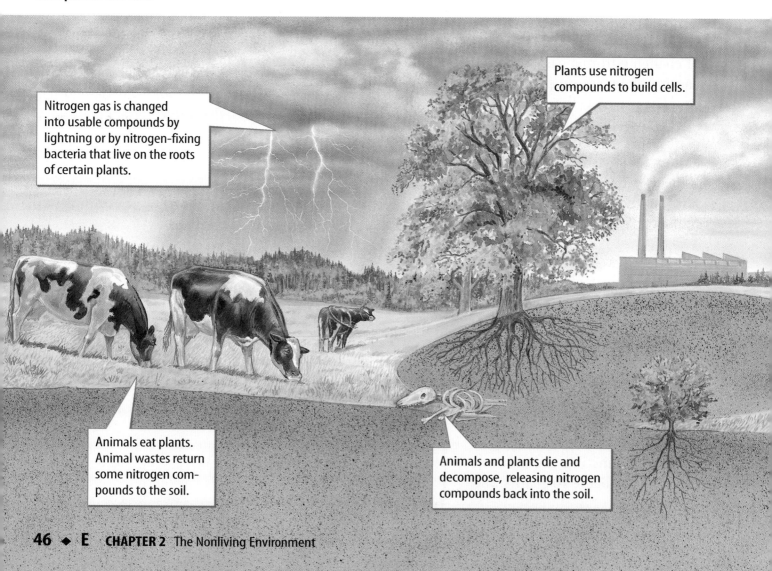

Nitrogen gas is changed into usable compounds by lightning or by nitrogen-fixing bacteria that live on the roots of certain plants.

Plants use nitrogen compounds to build cells.

Animals eat plants. Animal wastes return some nitrogen compounds to the soil.

Animals and plants die and decompose, releasing nitrogen compounds back into the soil.

Figure 12
Nitrogen fixation is important to plant growth.

A Soybeans can help restore nitrogen to the soil.

B The swollen nodules on the roots of the soybean plants contain colonies of nitrogen-fixing bacteria.

C The bacteria depend on the plant for food. The plant depends on the bacteria to form the nitrogen compounds the plant needs.

Magnification: 1,000×

Soil Nitrogen Human activities can affect the part of the nitrogen cycle that takes place in the soil. If a farmer grows a crop, such as corn or wheat, most of the plant material is taken away when the crop is harvested. The plants are not left in the field to decay and return their nitrogen compounds to the soil. If these nitrogen compounds are not replaced, the soil could become infertile. You might have noticed that adding fertilizer to soil can make plants grow greener, bushier, or taller. Most fertilizers contain the kinds of nitrogen compounds that plants need for growth. Fertilizers can be used to replace soil nitrogen in crop fields, lawns, and gardens. Compost and animal manure also contain nitrogen compounds that plants can use. They also can be added to soil to improve fertility.

Another method farmers use to replace soil nitrogen is to grow nitrogen-fixing crops. Most nitrogen-fixing bacteria live on or in the roots of certain plants. Some plants, such as peas, clover, and beans including the soybeans shown in **Figure 12,** have roots with swollen nodules that contain nitrogen-fixing bacteria. These bacteria supply nitrogen compounds to the soybean plants and add nitrogen compounds to the soil.

Mini LAB

Comparing Fertilizers

Procedure

1. Examine the three numbers (e.g., 5-10-5) on the **labels of three brands of house-plant fertilizer.** The numbers indicate the percentages of nitrogen, phosphorus, and potassium, respectively, that the product contains.
2. Compare the prices of the three brands of fertilizer.
3. Compare the amount of each brand needed to fertilize a typical houseplant.

Analysis

1. Which brand has the highest percentage of nitrogen?
2. Which brand is the most expensive source of nitrogen? The least expensive?

GEOGRAPHIC VISUALIZING THE CARBON CYCLE

Figure 13

Carbon—in the form of different kinds of carbon-containing molecules—moves through an endless cycle. The diagram below shows several stages of the carbon cycle. It begins when plants and algae remove carbon from the environment during photosynthesis. This carbon returns to the atmosphere via several carbon-cycle pathways.

Ⓐ Air contains carbon in the form of carbon dioxide gas. Plants and algae use carbon dioxide to make sugars, which are energy-rich, carbon-containing compounds.

Ⓑ Organisms break down sugar molecules made by plants and algae to obtain energy for life and growth. Carbon dioxide is released as a waste.

Ⓒ Burning fossil fuels and wood releases carbon dioxide into the atmosphere.

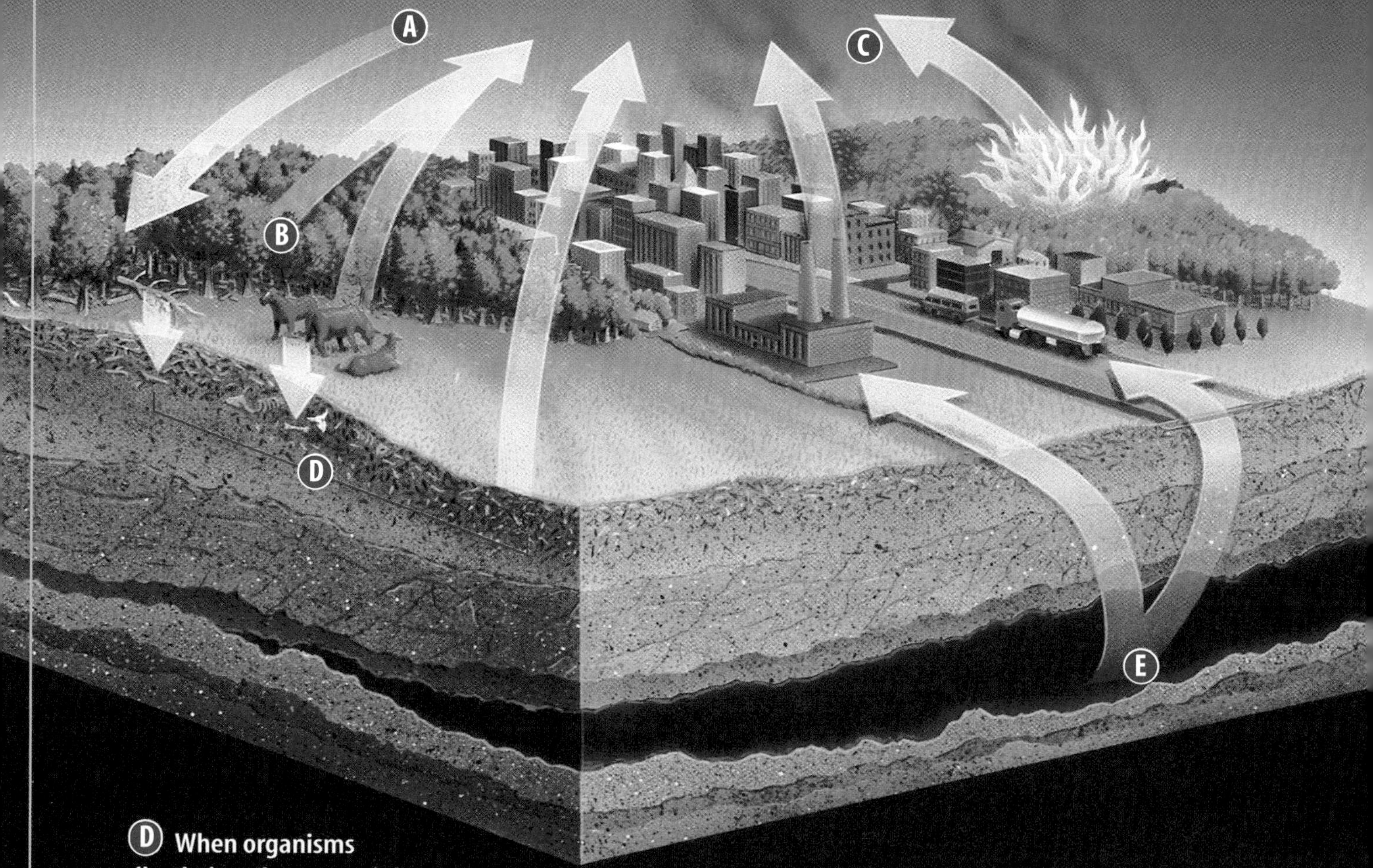

Ⓓ When organisms die, their carbon-containing molecules become part of the soil. The molecules are broken down by fungi, bacteria, and other decom-

Ⓔ Under certain conditions, the remains of some dead organisms may gradually be changed into fossil fuels such as coal, gas, and oil. These carbon

The Carbon Cycle

Carbon atoms are found in the molecules that make up living organisms. Carbon is an important part of soil humus, which is formed when dead organisms decay, and it is found in the atmosphere as carbon dioxide gas (CO_2). The **carbon cycle** describes how carbon molecules move between the living and nonliving world, as shown in **Figure 13.**

The carbon cycle begins when producers remove CO_2 from the air during photosynthesis. They use CO_2, water, and sunlight to produce energy-rich sugar molecules. Energy is released from these molecules during respiration—the chemical process that provides energy for cells. Respiration uses oxygen and releases CO_2. Photosynthesis uses CO_2 and releases oxygen. These two processes help recycle carbon on Earth.

Reading Check *How does carbon dioxide enter the atmosphere?*

Human activities also release CO_2 into the atmosphere. Fossil fuels such as gasoline, coal, and heating oil are the remains of organisms that lived millions of years ago. These fuels are made of energy-rich, carbon-based molecules. When people burn these fuels, CO_2 is released into the atmosphere as a waste product. People also use wood for building and for fuel. Trees that are harvested for these purposes no longer remove CO_2 from the atmosphere during photosynthesis. The amount of CO_2 in the atmosphere is increasing. Extra CO_2 could trap more heat from the Sun and cause average temperatures on Earth to rise.

SCIENCE *Online*

Research Visit the Glencoe Science Web site at **science.glencoe.com** for the chemical equations that describe photosynthesis and respiration. In your Science Journal, write these equations and use them to explain how respiration is the reverse of photosynthesis.

Section 2 Assessment

1. Describe the water cycle.
2. Explain how respiration can be considered the reverse of photosynthesis.
3. How might burning fossil fuels affect the composition of gases in the atmosphere?
4. Why do plants, animals, and other organisms need nitrogen?
5. **Think Critically** Most chemical fertilizers contain nitrogen, phosphorus, and potassium. Why don't they contain carbon? How do plants obtain carbon?

Skill Builder Activities

6. **Identifying and Manipulating Variables and Controls** Describe an experiment that would determine whether extra carbon dioxide enhances the growth of tomato plants. **For more help,** refer to the Science Skill Handbook.
7. **Communicating** Pretend you are a carbon molecule. Write a fictional account of your travels from the atmosphere, through at least two organisms, and back to the atmosphere. **For more help,** refer to the Science Skill Handbook.

Energy Flow

Figure 14
A Chemicals in the water that flows from hydrothermal vents provide bacteria with a source of energy. **B** The bacterial producers use this energy to make nutrients through the process of chemosynthesis. Consumers, such as tubeworms, feed on the bacteria.

Converting Energy

All living things are made of matter, and all living things need energy. Matter and energy move through the natural world in different ways. Matter can be recycled over and over again. The recycling of matter requires energy. Energy is not recycled, but it is converted from one form to another. The conversion of energy is important to all life on Earth.

Photosynthesis During photosynthesis, producers convert light energy into the chemical energy in sugar molecules. Some of these sugar molecules are broken down as energy is needed. Others are used to build complex carbohydrate molecules that become part of the producer's body. Fats and proteins also contain stored energy.

Chemosynthesis Not all producers rely on light for energy. During the 1970s, scientists exploring the ocean floor were amazed to find communities teeming with life. These communities were at a depth of almost 3.2 km and living in total darkness. They were found near powerful hydrothermal vents like the one shown in **Figure 14**.

A

B

Magnification: 38,000×

Hydrothermal Vents A hydrothermal vent is a deep crack in the ocean floor through which the heat of molten magma can escape. The water from hydrothermal vents is extremely hot from contact with molten rock that lies deep in Earth's crust.

Because no sunlight reaches these deep ocean regions, plants or algae cannot grow there. How do the organisms living in this community obtain energy? Scientists learned that the hot water contains nutrients such as sulfur molecules that bacteria use to produce their own food. The production of energy-rich nutrient molecules from chemicals is called **chemosynthesis** (kee moh SIN thuh sus). Consumers living in the hydrothermal vent communities rely on chemosynthetic bacteria for nutrients and energy. Chemosynthesis and photosynthesis allow producers to make their own energy-rich molecules.

✔ Reading Check *What is chemosynthesis?*

Energy Transfer

Energy can be converted from one form to another. It also can be transferred from one organism to another. Consumers cannot make their own food. Instead, they obtain energy by eating producers or other consumers. This way, energy stored in the molecules of one organism is transferred to another organism. At the same time, the matter that makes up those molecules is transferred from one organism to another. Throughout nature, energy and matter move from organism to organism when one organism becomes food for another organism.

Food Chains A food chain is a way of showing how matter and energy pass from one organism to another. Producers—plants, algae, and other organisms that are capable of photosynthesis or chemosynthesis—are always the first step in a food chain. Animals that consume producers such as herbivores are the second step. Carnivores and omnivores—animals that eat other consumers—are the third and higher steps of food chains. One example of a food chain is shown in **Figure 15.**

Earth Science
INTEGRATION

The first hydrothermal vent community discovered was found along the Galápagos rift zone. A rift zone forms where two plates of Earth's crust are spreading apart. In your Science Journal, describe the energy source that heats the water in the hydrothermal vents of the Galápagos rift zone.

Figure 15
In this food chain, grasses are producers, marmots are herbivores that eat the grasses, and grizzly bears are consumers that eat marmots. The arrows show the direction in which matter and energy flow.

Food Webs A forest community includes many feeding relationships. These relationships can be too complex to show with a food chain. For example, grizzly bears eat many different organisms, including berries, insects, chipmunks, and fish. Berries are eaten by bears, birds, insects, and other animals. A bear carcass might be eaten by wolves, birds, or insects. A **food web** is a model that shows all the possible feeding relationships among the organisms in a community. A food web is made up of many different food chains, as shown in **Figure 16.**

Energy Pyramids

Food chains usually have at least three links, but rarely more than five. This limit exists because the amount of available energy is reduced as you move from one level to the next in a food chain. Imagine a grass plant that absorbs energy from the Sun. The plant uses some of this energy to grow and produce seeds. Some of the energy is stored in the seeds.

Figure 16
Compared to a food chain, a food web provides a more complete model of the feeding relationships in a community.

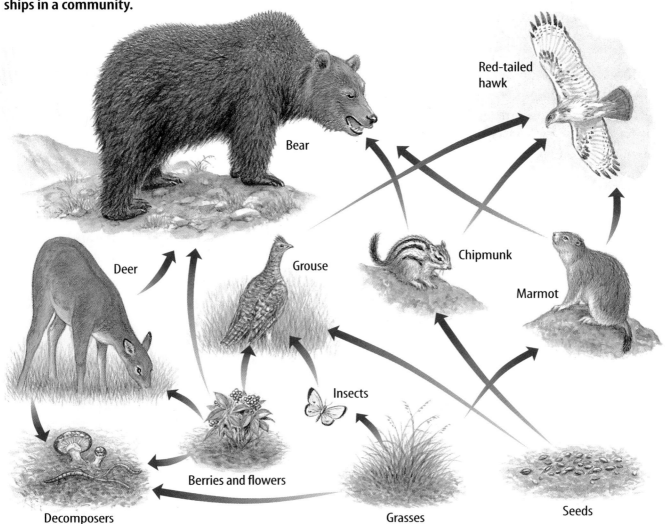

Bear

Red-tailed hawk

Chipmunk

Marmot

Deer

Grouse

Insects

Berries and flowers

Decomposers

Grasses

Seeds

Available Energy When a mouse eats grass seeds, energy stored in the seeds is transferred to the mouse. However, most of the energy the plant absorbed from the Sun was used for the plant's growth. Much less energy is stored in the seeds eaten by the mouse. The mouse uses much of the energy remaining in the seeds for its own life processes, including respiration, digestion, and growth. A hawk that eats the mouse obtains even less energy.

The same thing happens at every feeding level of a food chain. The amount of available energy is reduced from one feeding level to another. An **energy pyramid,** like the one in **Figure 17,** shows the amount of energy available at each feeding level in an ecosystem. The bottom layer of the pyramid, which represents all of the producers, is the first feeding level. It is the largest level because it contains the most energy and the largest number of organisms. As you move up the pyramid, each level becomes smaller. Only about ten percent of the energy available at each feeding level of an energy pyramid is transferred to the next higher level.

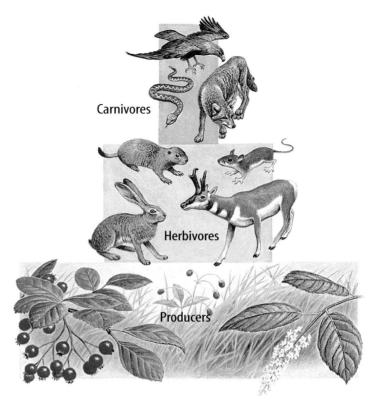

Carnivores

Herbivores

Producers

Figure 17
This energy pyramid shows that each feeding level contains less energy than the level below it. *What would happen if the hawks and snakes outnumbered the rabbits and mice in this ecosystem?*

 **Reading Check** *Why does the first feeding level of an energy pyramid contain the most energy?*

Section ③ Assessment

1. Compare and contrast photosynthesis and chemosynthesis.
2. Explain how your three favorite foods provide you with energy from the Sun.
3. What is the difference between a food web and an energy pyramid?
4. Why is there a limit to the number of links in a food chain?
5. **Think Critically** Use your knowledge of food chains and the energy pyramid to explain why the number of mice in a grassland ecosystem is greater than the number of hawks.

Skill Builder Activities

6. **Classifying** Classify each species as photosynthetic or chemosynthetic: *Red hattus* uses red light to make its food; *Selen dion* makes food if the element selenium is present. **For more help, refer to the** Science Skill Handbook.

7. **Solving One-Step Equations** A forest has 24,055,000 kilocalories (kcals) of producers, 2,515,000 kcals of herbivores, and 235,000 kcals of carnivores. How much energy is lost between producers and herbivores? Between herbivores and carnivores? **For more help, refer to the** Math Skill Handbook.

Activity

Where does the mass of a plant come from?

An enormous oak tree starts out as a tiny acorn. The acorn sprouts in dark, moist soil. Roots grow down through the soil. Its stem and leaves grow up toward the light and air. Year after year, the tree grows taller, its trunk grows thicker, and its roots grow deeper. It becomes a towering oak that produces thousands of acorns of its own. An oak tree has much more mass than an acorn. Where does this mass come from? The soil? The air? In this activity, you'll find out by conducting an experiment with radish plants.

What You'll Investigate

Does all of the matter in a radish plant come from the soil?

Goals

- **Measure** the mass of soil before and after radish plants have been grown in it.
- **Measure** the mass of radish plants grown in the soil.
- **Analyze** the data to determine whether the mass gained by the plants equals the mass lost by the soil.

Materials

8-oz plastic or paper cup
potting soil to fill cup
scale or balance
radish seeds (4)
water
paper towels

Safety Precautions 🥽 👕

Procedure

1. Copy the data table into your Science Journal.
2. Fill the cup with dry soil.
3. Find the mass of the cup of soil and record this value in your data table.
4. Moisten the soil in the cup. Plant four radish seeds 2 cm deep in the soil. Space the seeds an equal distance apart. Wash your hands.
5. Add water to keep the soil barely moist as the seeds sprout and grow.
6. When the plants have developed four to six true leaves, usually after two to three weeks, carefully remove the plants from the soil. Gently brush the soil off the roots. Make sure all the soil remains in the cup.
7. Spread the plants out on a paper towel. Place the plants and the cup of soil in a warm area to dry out.

8. When the plants are dry, measure their mass and record this value in your data table. Write this number with a plus sign in the Gain or Loss column.
9. When the soil is dry, find the mass of the cup of soil. Record this value in your data table. Subtract the End mass from the Start mass and record this number with a minus sign in the Gain or Loss column.

Mass of Soil and Radish Plants			
	Start	End	Gain (+) or Loss (−)
Mass of dry soil and cup			
Mass of dried radish plants	0 g		

Conclude and Apply

1. In the early 1600s, a Belgian scientist named J. B. van Helmont conducted this experiment with a willow tree. What is the advantage of using radishes instead of a tree?
2. How much mass was gained or lost by the soil? By the radish plants?
3. Did the mass of the plants come completely from the soil? How do you know?

4. If all of the mass gained by the plants did not come from the soil, where could it have come from?

Communicating Your Data

Compare your conclusions with those of other students in your class. **For more help, refer to the** Science Skill Handbook.

Extreme Climates

Did you know...

... The greatest snowfall in one year occurred at Mount Baker in Washington State. Approximately 2,896 cm of snow fell on Mount Baker during the 1998-99, 12-month snowfall season. That's enough snow to bury an eight-story building.

2,896 cm

... The hottest climate in the United States is found in Death Valley, California. In July 1913, Death Valley reached approximately 57°C. This is the hottest officially recognized temperature on Earth. As a comparison, a comfortable room temperature is about 20°C.

... The record for the lowest temperature was set in Antarctica in 1983. The temperature was a frigid −89°C. As a comparison, the temperature of your freezer at home is about −15°C.

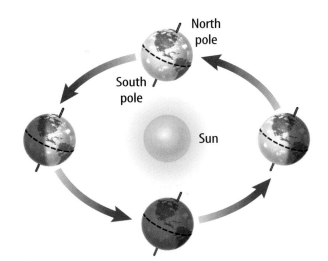

North pole

South pole

Sun

... The south pole receives sunshine for less than 50 percent of the days in a year. Because Earth is tilted, the south pole is pointed away from the Sun for about half the year and receives very little during that time.

... The fastest tornado winds have been measured at a speed of about 512 km/h. That's faster than the blades of some helicopters, which can rotate at about 450 km/h.

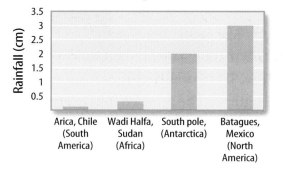

Lowest Average Annual Rainfall

Rainfall (cm)

3.5
3
2.5
2
1.5
1
0.5

Arica, Chile (South America) | Wadi Halfa, Sudan (Africa) | South pole (Antarctica) | Batagues, Mexico (North America)

Highest Average Annual Rainfall

Rainfall (cm)

1400
1050
700
350

Lloro, Colombia (South America) | Mawsynram, India (Asia) | Mount Waialeale, Hawaii (Pacific Ocean) | Debundscha, Cameroon (Africa)

Do the Math

1. Look at the graph above. How many years of average south pole precipitation would it take to equal a single year of average precipitation in Lloro, Colombia?
2. What is the difference in degrees Celsius between the world record low temperature and the world record high temperature?
3. What was the average monthly snowfall at Mount Baker during the 1998-99 snowfall season?

Go Further

Go to **science.glencoe.com** and find out the average monthly rainfall in a tropical rain forest. Make a line graph to show how the amount of precipitation changes during the 12 months of the year.

Chapter ② Study Guide

Reviewing Main Ideas

Section 1 Abiotic Factors

1. Abiotic factors include air, water, soil, sunlight, temperature, and climate. *What abiotic factors are required for this squirrel's survival? Explain.*

2. The availability of water and light influences where life exists on Earth.

3. Soil and climate have an important influence on the types of organisms that can survive in different environments.

4. High latitudes and elevations generally have lower average temperatures.

Section 2 Cycles in Nature

1. Matter is limited on Earth and is recycled through the environment. *How do green plants help recycle oxygen?*

2. The water cycle involves evaporation, condensation, and precipitation.

3. The carbon cycle involves photosynthesis and respiration.

4. Nitrogen in the form of soil compounds enters plants, which are then consumed by other organisms.

Section 3 Energy Flow

1. Producers make energy-rich molecules through photosynthesis or chemosynthesis. *How do seaweeds in shallow water obtain energy?*

2. When organisms feed on other organisms, they obtain matter and energy.

3. Matter can be recycled, but energy cannot.

4. Food webs are models of the complex feeding relationships in communities.

5. Available energy decreases as you go to higher feeding levels in an energy pyramid. *What happens to most of the energy in an apple that you eat?*

FOLDABLES
Reading & Study Skills

After You Read

Find a student who drew a different ecosystem on his or her Cause and Effect Study Fold. Then, compare and contrast the information on your two Foldables.

Visualizing Main Ideas

This diagram shows photosynthesis in a leaf. Fill in the blank lines with the terms light, carbon dioxide, *and* oxygen.

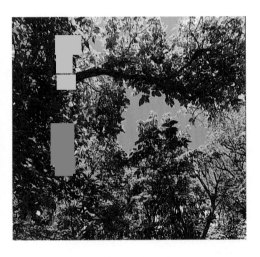

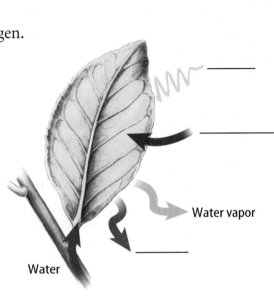

Water vapor

Water

Vocabulary Review

Vocabulary Words

a. abiotic
b. atmosphere
c. biotic
d. carbon cycle
e. chemosynthesis
f. climate
g. condensation
h. energy pyramid
i. evaporation
j. food web
k. nitrogen cycle
l. nitrogen fixation
m. soil
n. water cycle

THE PRINCETON REVIEW | **Study Tip**

Write out the full questions and answers to end-of-chapter quizzes, not just the answers. This will help you form complete responses to important questions.

Using Vocabulary

Which vocabulary word best corresponds to each of the following events?

1. A liquid changes to a gas.

2. Some types of bacteria form nitrogen compounds in the soil.

3. Decaying plants add nitrogen to the soil.

4. Chemical energy is used to make energy-rich molecules.

5. Decaying plants add carbon to the soil.

6. A gas changes to a liquid.

7. Water flows downhill into a stream. The stream flows into a lake, and water evaporates from the lake.

8. Burning coal and exhaust from automobiles release carbon into the air.

Chapter 2 Assessment

Checking Concepts

Choose the word or phrase that best answers the question.

1. Which of the following is an abiotic factor?
   A) penguins   C) soil bacteria
   B) rain   D) redwood trees

2. Which group makes up the largest level of an energy pyramid?
   A) herbivores   C) decomposers
   B) producers   D) carnivores

3. You climb up the western slope of the Cascade Mountains and down the eastern side. Which of the following weather changes do you observe?
   A) Warm and wet changes to cold and wet, then cold and dry, then warm and dry.
   B) Cold and wet changes to warm and wet, then warm and dry, then cold and dry.
   C) Warm and wet changes to cold and wet, then cold and dry, then warm and wet.
   D) Warm and dry changes to cold and dry, then warm and dry, then cold and dry.

4. Which of the following applies to latitudes farther from the equator?
   A) higher elevations
   B) higher temperatures
   C) higher precipitation levels
   D) lower temperatures

5. Water vapor forming droplets that form clouds directly involves which process?
   A) condensation   C) evaporation
   B) respiration   D) transpiration

6. Which one of the following components of air is least necessary for life on Earth?
   A) argon   C) carbon dioxide
   B) nitrogen   D) oxygen

7. What do plants make that requires nitrogen?
   A) sugars   C) fats
   B) proteins   D) carbohydrates

8. Which of the following processes removes carbon dioxide from the air?
   A) condensation   C) burning
   B) photosynthesis   D) respiration

9. Earth receives a constant supply of which of the following items?
   A) light energy   C) nitrogen
   B) carbon   D) water

10. Which of these is an energy source for chemosynthesis?
    A) sunlight   C) sulfur molecules
    B) moonlight   D) carnivores

Thinking Critically

11. A country has many starving people. Should they grow vegetables and corn to eat, or should they grow corn to feed cattle so they can eat beef? Explain.

12. Why is a food web a better model than a food chain?

13. Do bacteria need nitrogen? Why or why not?

14. It is often easier to walk through an old, mature forest of tall trees than through a young forest that is full of small trees. Why?

15. The Inyo Mountains are located in central California. Explain why giant sequoia trees grow on the west side of the mountains and Death Valley, a desert, is on the east side.

Developing Skills

16. **Classifying** Classify each of the following environmental concerns according to the cycle it affects—carbon, nitrogen, or water.
    a. algal blooms caused by excess fertilizer
    b. acid rain damage to pine trees
    c. the unnatural warming of Earth

17. Recognizing Cause and Effect A lake in Kenya has been taken over by a floating weed. What could you do to determine if nitrogen fertilizer runoff from farms is causing the problem?

18. Making and Using Graphs Abiotic factors, such as climate, cause populations to move from place to place. Make a bar graph of the following migration distances.

Mighty Migrators	
Species	**Distance (km)**
Desert locust	4,800
Caribou	800
Green turtle	1,900
Arctic tern	35,000
Gray whale	19,000

19. Forming Hypotheses For each hectare of land, ecologists found 10,000 kcals of producers, 10,000 kcals of herbivores, and 2,000 kcals of carnivores. Suggest a reason why producer and herbivore levels are equal.

20. Concept Mapping Draw a food web of these organisms: *caterpillars and rabbits eat grasses, raccoons eat rabbits and mice, mice eat grass seeds,* and *birds eat caterpillars.*

Performance Assessment

21. Poster Use magazine photographs to make a visual representation of the water cycle.

THE PRINCETON REVIEW **Test Practice**

Food chains model how energy is transferred from one organism to another organism in the environment. The diagram below shows a food chain that includes aquatic plants and animals and a land animal.

Study the diagram and answer the following questions.

1. Which of the following statements is true based on the order of the food chain shown above?
 A) Algae eat plankton.
 B) Plankton eat salmon.
 C) Herring eat plankton.
 D) Salmon eat algae.

2. If the supply of salmon were suddenly depleted, the numbers of which of the following might also be depleted?
 F) algae   **H)** herring
 G) plankton   **J)** eagles

3 Ecosystems

During the summer of 2000, wildfires burned out of control in Montana and other western states. Can this Montana hillside recover from the forest fire? Can the animals that lived here before the fire adapt to the changed environment? How is the forest in this picture different from a rain forest in South America? In this chapter, you will learn about the different ecosystems that exist on Earth. You also will learn how ecosystems change and how plants and animals adapt to those changes.

What do you think?

Science Journal Look at the picture below with a classmate. Discuss what this might be or what is happening. Here's a hint: *It is the first step in creating land.* Write your answer or best guess in your Science Journal.

Ehe plants growing in your classroom or home may not look like the same types of plants that you find growing outside. Where do indoor plants come from? Many indoor plants don't grow well outside in most North American climates. Do the activity below to determine what type of climate most houseplants thrive in.

Infer the origin of houseplants

1. Examine a healthy houseplant in your classroom or home.

2. Describe the environmental conditions found in your classroom or home. For example, is the air humid or dry? Are temperatures warm and nearly constant?

3. Using observations from step 1 and descriptions from step 2, hypothesize about the natural environment of the plants in your classroom or home.

Observe

In your Science Journal, record the observations that led to your hypothesis. How would you design an experiment to test your hypothesis?

Before You Read

FOLDABLES
Reading & Study Skills

Making a Main Ideas Study Fold A main idea consists of the major concepts or topics talked about in a chapter. Before you read the chapter, make the following Foldable to help you understand succession, or the gradual change in the types of species that live in an area.

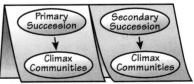

1. Place a sheet of paper in front of you so the long side is at the top. Fold the paper in half from the left side to the right side. Fold top to bottom but do not crease and unfold.

2. Turn the paper so the fold is at the top and label *Primary Succession* and *Secondary Succession* across the front of the paper, as shown. Draw an arrow down from each and then label *Climax Communities,* as shown. Circle all four labels.

3. Through the top thickness of paper, cut along the middle fold line to form two tabs as shown.

4. As you read the chapter, define terms and collect information under the tabs.

How Ecosystems Change

As You Read

***What* You'll Learn**

- **Explain** how ecosystems change over time.
- **Describe** how new communities begin in areas without life.
- **Compare** pioneer species and climax communities.

Vocabulary

succession
pioneer species
climax community

***Why* It's Important**

Understanding how ecosystems change can help you predict what can happen to the land around you in the years to come.

Ecological Succession

What would happen if the lawn at your home were never cut? The grass would get longer, as in **Figure 1,** and soon it would look like a meadow. Later, larger plants would grow from seeds brought to the area by animals or wind. Then, trees might sprout. In fact, in 20 years or less you wouldn't be able to tell that the land was once a mowed lawn. An ecologist can tell you what type of ecosystem your lawn would become. If it would become a forest, they can tell you how long it would take and predict the type of trees that would grow there. **Succession** refers to the normal, gradual changes that occur in the types of species that live in an area. Succession occurs differently in different places around the world.

Primary Succession As lava flows from the mouth of a volcano, it is so hot that it destroys everything in its path. When it cools, lava forms new land composed of rock. It is hard to imagine that this land eventually could become a forest or grassland someday.

The process of succession that begins in a place without any soil is called primary succession. It starts with the arrival of living things such as lichens (LI kunz). These living things, called **pioneer species,** do not need soil to survive. They survive drought, extreme heat and cold, and other harsh conditions and start the soil-building process.

Figure 1
Open areas that are not maintained will become overgrown with grasses and shrubs as succession begins.

Figure 2
Lichens, like these in Colorado, are fragile and take many years to grow. They often cling to bare rock where many other organisms can't survive. *How do lichens form soil?*

New Soil During primary succession, shown in **Figure 2,** soil begins to form as lichens and the forces of weather and erosion help break down rocks into smaller pieces. When lichens die, they decay, adding small amounts of organic matter to the rock. Plants such as mosses and ferns can grow in this new soil. Eventually, these plants die, adding more organic material. The soil layer thickens, and grasses, wildflowers, and other plants begin to take over. When these plants die, they add more nutrients to the soil. This buildup is enough to support the growth of shrubs and trees. All the while, insects, small birds, and mammals have begun to move in. What was once bare rock now supports a variety of life.

Secondary Succession What happens when a fire, such as the one in **Figure 3,** destroys a forest or when a building is torn down in a city? After a forest fire, not much is left except dead trees and ash-covered soil. After the rubble of a building is removed, all that remains is bare soil. However, these places do not remain lifeless for long. The soil already contains the seeds of weeds, grasses, and trees. More seeds are carried to the area by wind and birds. Other wildlife may move in. Succession that begins in a place that already has soil and was once the home of living organisms is called secondary succession. Because soil already is present, secondary succession occurs faster and has different pioneer species than primary succession does.

SCIENCE *Online*

Data Update Visit the Glencoe Science Web site at **science.glencoe.com** to learn how forests in the western United States are recovering from the wildfires of 2000. Communicate to your class what you learn.

✔ **Reading Check** *Which type of succession starts without soil?*

Figure 3

In the summer of 1988, wind-driven flames like those shown in the background photo swept through Yellowstone National Park, scorching nearly a million acres. The Yellowstone fire was one of the worst forest fires in United States history. The images on this page show secondary succession—the process of ecological regeneration—triggered by the fire.

▶ After the fire, burned timber and blackened soil seemed to be all that remained. However, the fire didn't destroy the seeds that were protected under the soil.

◀ Within weeks, grasses and other plants were beginning to grow in the burned areas. Ecological succession was underway.

▶ Many burned areas in the park opened new plots for stands of trees. This picture shows young lodgepole pines in August 1999. The forest habitat of America's oldest national park is being restored gradually through secondary succession.

Figure 4
This beech-maple forest is an example of a climax community.

Climax Communities A community that has reached a stable stage of ecological succession is called a **climax community.** It is a combination of plants and animals that use the available resources most efficiently. The beech-maple forest shown in **Figure 4** is an example of a community that has reached the end of succession. Diversity and balance are maintained in a climax community because as trees die, they provide nutrients for new communities of organisms. Some physical disturbances also are important in maintaining equilibrium in a climax community.

Primary succession begins in areas with no life at all. It can take hundreds or even thousands of years to develop into a climax community. Secondary succession is usually a shorter process, but it still can take a century or more.

Section 1 Assessment

1. What is ecological succession?
2. Explain the difference between primary and secondary succession.
3. What is the difference between pioneer species and climax communities?
4. What kind of succession will take place on an abandoned, unpaved country road?
5. **Think Critically** After a climax community is established in an area, explain how long it can survive there.

Skill Builder Activities

6. **Concept Mapping** Show the sequence of events in primary succession. Include the term *climax community.* **For more help, refer to the Science Skill Handbook.**

7. **Solving One-Step Equations** A tombstone etched with 1802 as the date of death has a lichen on it that is 6 cm in diameter. Assuming that the lichen started to grow in 1802, calculate the average yearly growth of the lichen. **For more help, refer to the Math Skill Handbook.**

Biomes

As You Read

What You'll Learn

- **Explain** how climate influences land environments.
- **Identify** seven biomes of Earth.
- **Describe** the adaptations of organisms found in each biome.

Vocabulary

biome   temperate deciduous forest
tundra   temperate rain forest
taiga   tropical rain forest
desert   grassland

Why It's Important

Resources that you need to survive are found in a variety of biomes.

Figure 5
The land portion of Earth can be divided into seven biomes.

Factors That Affect Biomes

Does a desert in Arizona have anything in common with a desert in Africa? Both have heat, little rain, poor soil, water-conserving plants with thorns, and lizards. Even widely separated regions of the world can have similar biomes because they have similar climates. Climate is the average weather pattern in an area over a long period of time. The two most important climatic factors that affect life in an area are temperature and precipitation.

Major Biomes

Large geographic areas that have similar climates and eco-systems are called **biomes** (BI ohmz). Seven common types of biomes are mapped in **Figure 5.** Areas with similar climates produce similar climax communities. Tropical rain forests are climax communities found near the equator, where temperatures are warm and rainfall is plentiful. Coniferous forests grow where winter temperatures are cold and rainfall is moderate.

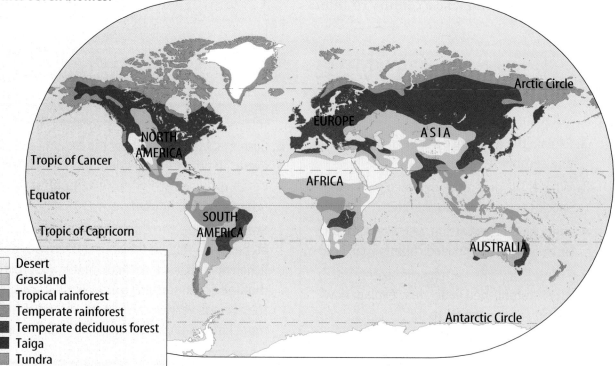

Tundra At latitudes just south of the north pole lies a biome that receives little precipitation but is covered with ice most of the year. The **tundra** is a cold, dry, treeless region, sometimes called a cold desert. Precipitation averages less than 25 cm per year. Winters can be six to nine months long. For some of these months, when the Sun never appears above the horizon, the land is dark 24 hours a day. The average temperature of the tundra is about −12°C. For a few days during the short, cold summer, the Sun is always visible. Only the top portion of soil thaws in the summer. Below the thawed surface is a layer of permanently frozen soil called permafrost, shown in **Figure 6.** Tundra soil has few nutrients because the cold temperatures slow the process of decomposition.

Figure 6
This permafrost in Alaska is covered by soil that freezes in the winter and thaws in the summer. *What types of problems might this cause for people living in this area?*

Tundra Life Tundra plants are adapted to drought and cold. They include mosses, grasses, and small shrubs, as seen in **Figure 7.** Many lichens grow on the tundra. During the summer, mosquitoes, blackflies, and other biting insects fill the air. Migratory birds such as ducks, geese, shorebirds, and songbirds nest on the tundra during the summer. Other inhabitants include hawks, snowy owls, and willow grouse. Mice, voles, lemmings, arctic hares, caribou, reindeer, and musk oxen also are found there.

People are concerned about overgrazing by animals on the tundra. Fences, roads, and pipelines have disrupted the migratory routes of some animals and forced them to stay in a limited area. Because the growing season is so short, plants and other vegetation can take decades to recover from damage.

Figure 7
Ⓐ Lichens, mosses, grasses, and small shrubs thrive on the tundra. Ⓑ Ptarmigan live on the tundra. In winter, their feathers turn white. Extra feathers on their feet keep them warm and prevent them from sinking into the snow.

Figure 8
A The taiga is dominated by cone-bearing evergreen trees.
B The lynx has broad, heavily furred feet that act like snowshoes to prevent it from sinking in the snow.

Taiga South of the tundra—between latitudes 50°N and 60°N and stretching across Canada, northern Europe, and Asia—is the world's largest biome. The **taiga** (TI guh), shown in **Figure 8,** is a cold, forest region dominated by cone-bearing evergreen trees. Although the winter is long and cold, the taiga is warmer and wetter than the tundra. Precipitation is mostly snow and averages 35 cm to 100 cm each year.

Most soils of the taiga thaw completely during the summer, making it possible for trees to grow. However, permafrost is present in the extreme northern regions of the taiga. The forests of the taiga might be so dense that little sunlight penetrates the trees to reach the forest floor. However, some lichens and mosses do grow on the forest floor. Moose, lynx, shrews, bears, and foxes are some of the animals that live in the taiga.

Temperate Deciduous Forest Temperate regions usually have four distinct seasons each year. Annual precipitation ranges from about 75 cm to 150 cm and is distributed throughout the year. Temperatures range from below freezing during the winter to 30°C or more during the warmest days of summer.

Figure 9
A The leaves on trees in deciduous forests change color in autumn.
B White-tailed deer are found in deciduous forests. They feed on a variety of plants and nuts.

Temperate Forest Life Many ever-green trees grow in the temperate regions of the world. However, most of the temperate forests in Europe and North America are dominated by climax communities of deciduous trees, which lose their leaves every autumn. These forests, like the one in **Figure 9,** are called **temperate deciduous forests.** In the United States, most of them are located east of the Mississippi River.

When European settlers first came to America, they cut trees to create farmland and to supply wood. As forests were cut, organisms lost their habitats. When agriculture shifted from the eastern to the midwestern and western states, secondary succession began, and trees eventually returned to some areas. Now, nearly as many trees grow in the New England states as did before the American Revolutionary War. Many trees are located in smaller patches. Yet, the recovery of large forests such as those in the Adirondack Mountains in New York State shows that succession is possible.

Temperate Rain Forest New Zealand, southern Chile, and the Pacific Northwest of the United States are some of the places where **temperate rain forests,** shown in **Figure 10,** are found. The average temperature of a temperate rain forest ranges from 9°C to 12°C. Precipitation ranges from 200 cm to 400 cm per year.

Trees with needlelike leaves dominate these forests, including the Douglas fir, western red cedar, and spruce. Many grow to great heights. Animals of the temperate rain forest include the black bear, cougar, bobcat, endangered northern spotted owl, and threatened marbled murrelet. Many species of amphibians also inhabit the temperate rain forest, including salamanders.

The logging industry in the Northwest provides jobs for many people. However, it also removes large parts of the temperate rain forest and destroys the habitat of many organisms. Many logging companies now are required to replant trees to replace the ones they cut down. Also, some rain forest areas are protected as national parks and forests.

Figure 10
A In the Olympic rain forest in Washington State, mosses and lichens blanket the ground and hang from the trees. **B** Wet areas are perfect habitats for amphibians like this Pacific giant salamander.

Figure 11
Tropical rain forests are lush environments that contain such a large variety of species that many have not been discovered.

TRY AT HOME

Modeling Rain Forest Leaves

Procedure

1. Draw an oval leaf about 10 cm long on a piece of **poster board.** Cut it out.
2. Draw a second leaf the same size but make the end pointed. This is called a drip tip. Cut this leaf out.
3. Lay a leaf on each of your hands over a **sink.** Point the drip tip away from you. Tilt your hands down but do not allow the leaves to fall off.
4. Have someone gently spray **water** on the leaves and observe what happens.

Analysis

1. From which leaf does water drain faster?
2. Infer why it is an advantage for a leaf to get rid of water quickly in a rain forest.

Tropical Rain Forest Warm temperatures, wet weather, and lush plant growth are found in **tropical rain forests.** These forests are warm because they are near the equator. The average temperature, about 25°C, doesn't vary much between night and day. Most tropical rain forests receive at least 200 cm of rain annually. Some receive as much as 600 cm of rain each year.

Tropical rain forests, like the one in **Figure 11,** are home to an astonishing variety of organisms. They are the most biologically diverse places in the world. For example, one tree in a South American rain forest might contain more species of ants than exist in all of the British Isles.

Tropical Rain Forest Life Different animals and plants live in different parts of the rain forest. Scientists divide the rain forest into zones based on the types of plants and animals that live there, just as a library separates books about different topics onto separate shelves. The zones are as follows: forest floor, understory, canopy, and emergents, as shown in **Figure 12.** These zones often blend together, but their existence allows many organisms to live in the tropical rain forest.

✔ **Reading Check** *What are the four zones of a tropical rain forest?*

Although tropical rain forests support a huge variety of organisms, the soil of the rain forest contains few nutrients. The soil is poor because of the high number of organisms that live in the soil and because of erosion by water. Any dead leaves or vegetation are consumed quickly.

Human Impact Farmers that live in tropical areas clear the land to farm and to sell the valuable wood. After a few years, the crops use up the nutrients in the soil, and the farmers must clear more land. As a result, tropical rain forest habitats are being destroyed. Through education, people are realizing the value and potential value of preserving the species of the rain forest. In some areas, logging is prohibited. In other areas, farmers are taught new methods of farming so they do not have to clear rain forest lands continually.

Is where you live like a tropical rain forest or some other biome? To find out more about biomes, see the Biome Field Guide at the back of the book.

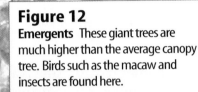

Figure 12
Emergents These giant trees are much higher than the average canopy tree. Birds such as the macaw and insects are found here.

Canopy The canopy includes the upper parts of the trees. It's full of life—insects, birds, reptiles, and mammals.

Understory This dark, cool environment is under the canopy leaves but above the ground. Many insects, reptiles, and amphibians live in the understory.

Forest Floor The forest floor is full of life. Many insects and the largest mammals in the rain forest generally live here.

When vegetation is removed from soil in areas that receive little rain, the dry, unprotected surface can be blown away. If the soil remains bare, a desert might form. This process is called desertification. Look on a biome map and hypothesize about which areas of the United States are most likely to become deserts.

Figure 13

A Desert plants, like these in the Sonoran Desert, are adapted for survival in the extreme conditions of the desert biome. **B** The giant hairy scorpion found in some deserts has a venomous sting.

Desert The driest biome on Earth is the **desert.** Deserts receive less than 25 cm of rain each year and support little plant life. Some desert areas receive no rain for years. When rain does come, it quickly drains away. Any water that remains on the ground evaporates rapidly.

Most deserts, like the one in **Figure 13,** are covered with a thin, sandy, or gravelly soil that contains little organic matter. Due to the lack of water, desert plants are spaced far apart and much of the ground is bare. Barren, windblown sand dunes are characteristics of the driest deserts.

✔ **Reading Check** *Why is much of a desert bare ground?*

Desert Life Desert plants are adapted for survival in the extreme dryness and hot and cold temperatures of this biome. Cactus plants are probably the most familiar desert plants of the western hemisphere. Desert animals also have adaptations that help them survive the extreme conditions. Some, like the kangaroo rat, never need to drink water. They get all the moisture they need from the breakdown of food during digestion. Most animals are active only during the night, late afternoon, or early morning when temperatures are less extreme. Few large animals are found in the desert.

In order to provide water for desert cities, rivers and streams have been diverted. When this happens, wildlife tends to move closer to cities in their search for food and water. Education about desert environments has led to an awareness of the impact of human activities. As a result, large areas of desert have been set aside as national parks and wilderness areas to protect desert habitats.

Grasslands Temperate and tropical regions that receive between 25 cm and 75 cm of precipitation each year and are dominated by climax communities of grasses are called **grasslands.** Most grasslands have a dry season, when little or no rain falls. This lack of moisture prevents the development of forests. Grasslands are found in many places around the world, and they have a variety of names. The prairies and plains of North America, the steppes of Asia, the savannas of Africa shown in **Figure 14,** and the pampas of South America are types of grasslands.

Figure 14
Animals such as zebras and wildebeests feed on the grasses of the savannas in Africa.

Grasslands Life The most noticeable animals in grassland biomes are usually mammals that graze on the stems, leaves, and seeds of grass plants. Kangaroos graze in the grasslands of Australia. In Africa, communities of animals such as wildebeests, impalas, and zebras thrive in the savannas.

Grasslands are perfect for growing many crops such as wheat, rye, oats, barley, and corn. Grasslands also are used to raise cattle and sheep. However, overgrazing can result in the death of grasses and the loss of valuable topsoil from erosion. Most farmers and ranchers take precautions to prevent the loss of valuable habitats and soil.

Section  Assessment

1. Which two biomes are the driest?
2. Compare and contrast tundra organisms and desert organisms.
3. What is the biggest climatic difference between a temperate rain forest and a tropical rain forest?
4. Why does the soil of tropical rain forests make poor farmland?
5. **Think Critically** If you climb a mountain in Arizona, you might reach an area where the trees resemble the taiga trees in northern Canada. Why would a taiga forest exist in Arizona?

Skill Builder Activities

6. **Recording Observations** Animals have adaptations that help them survive in their environments. Make a list of animals that live in your area, and record the physical or behavioral adaptations that help them survive. **For more help, refer to the** Science Skill Handbook.

7. **Using a Database** Create a database of information about Earth's biomes. Include data about *temperature range, annual precipitation, limiting factors,* and *descriptions of climax communities.* **For more help, refer to the Technology Skill Handbook.**

Activity

Studying a Land Ecosystem

An ecological study includes observation and analysis of organisms and the physical features of the environment.

What You'll Investigate
How do you study an ecosystem?

Materials
graph paper	binoculars
thermometer	pencil
hand lens	field guides
notebook	compass

Goals
- **Observe** biotic factors and abiotic factors of an ecosystem.
- **Analyze** the relationships among organisms and their environment.

Safety Precautions

Environmental Observations		
Date		
Time of Day		
Temperature		
Organisms Observed		
Comments		

Procedure

1. Choose a portion of an ecosystem near your school or home to study. You might choose a decaying log, a pond, a forest area in a park, a garden, or even a crack in the sidewalk.

2. Determine the boundaries of your study area.

3. Using a tape measure and graph paper, make a map of your area. Determine north.

4. **Record** your observations in a table similar to the one shown on this page.

5. **Observe** the organisms in your study area. Use field guides to identify them. Use a hand lens to study small organisms and binoculars to study animals you can't get near. Look for evidence (such as tracks or feathers) of organisms you do not see.

6. Use a thermometer to measure the air temperature in your study area. Record this information in the table.

7. Visit your study area as many times as you can and at different times of the day for one week. At each visit, make the same measurements and record all observations. Note how the living and nonliving parts of the ecosystem interact.

Conclude and Apply

1. **Predict** what might happen if one or more abiotic factors were changed suddenly.

2. What might happen if one or more populations of plants or animals were removed from the area?

3. Form a hypothesis about the effect that a new population of organisms might have on your ecosystem.

Communicating Your Data

Make a poster of your data. Then compare your data with those of other students in your class. **For more help, refer to the** Science Skill Handbook.

Aquatic Ecosystems

Freshwater Ecosystems

In a land environment, temperature and precipitation are the most important factors that determine which species can survive. In aquatic environments, water temperature, the amount of sunlight present, and the amounts of dissolved oxygen and salt in the water are important. Earth's freshwater ecosystems include flowing water such as rivers and streams and standing water such as lakes, ponds, and wetlands.

Rivers and Streams Flowing freshwater environments vary from small, gurgling brooks to large, slow-moving rivers. Currents can quickly wash loose particles downstream, leaving a rocky or gravelly bottom. As the water tumbles and splashes, as shown in **Figure 15,** air from the atmosphere mixes in. Naturally fast-flowing streams usually have clearer water and higher oxygen content than slow-flowing streams.

Most nutrients that support life in flowing-water ecosystems are washed into the water from land. In areas where the water movement slows, such as in the pools of streams or in large rivers, debris settles to the bottom. These environments tend to have higher nutrient levels and more plant growth. They contain organisms that are not as well adapted to swiftly flowing water, such as freshwater mussels, minnows, and leeches.

As You Read

***What* You'll Learn**
- **Compare** flowing freshwater and standing freshwater ecosystems.
- **Identify** and describe important saltwater ecosystems.
- **Identify** problems that affect aquatic ecosystems.

Vocabulary

wetland	intertidal zone
coral reef	estuary

***Why* It's Important**
All of the life processes in your body depend on water.

Figure 15
Streams like this one are high in oxygen because of the swift, tumbling water. *Where do most nutrients in streams come from?*

Modeling Freshwater Environments

Procedure

1. Obtain a sample of **pond sediment or debris, plants, water, and organisms** from your teacher.
2. Cover the bottom of a **clear, plastic container** with about 2 cm of the debris.
3. Add one or two plants to the container.
4. Carefully pour pond water into the container until it is about two-thirds full.
5. Use a **net** to add several organisms to the water. Seal the container.
6. Using a **hand lens,** observe as many organisms as possible. Record your observations. Return your sample to its original habitat.

Analysis

Write a short paragraph describing the organisms in your sample. How did the organisms interact with each other?

Human Impact People use rivers and streams for many activities. Once regarded as a free place to dump sewage and other pollutants, many people now recognize the damage this causes. Treating sewage and restricting pollutants have led to an improvement in the water quality in some rivers.

Lakes and Ponds When a low place in the land fills with rainwater, snowmelt, or water from an overflowing stream, a lake or pond might form. Pond or lake water hardly moves. It contains more plant growth than flowing-water environments contain.

Lakes, such as the one in **Figure 16A,** are larger and deeper than ponds. They have more open water because most plant growth is limited to shallow areas along the shoreline. In fact, organisms found in the warm, sunlit waters of the shorelines often are similar to those found in ponds. If you were to dive to the bottom, you would discover few, if any, plants or algae growing. Colder temperatures and lower light levels limit the types of organisms that can live in deep lake waters. Floating in the warm, sunlit waters near the surface of freshwater lakes and ponds are microscopic algae, plants, and other organisms known as plankton.

A pond, shown in **Figure 16B,** is a small, shallow body of water. Because ponds are shallow, they are filled with animal and plant life. Sunlight usually penetrates to the bottom. The warm, sunlit water promotes the growth of plants and algae. In fact, many ponds are filled almost completely with plant material, so the only clear, open water is at the center. Because of the lush growth in pond environments, they tend to be high in nutrients.

Figure 16
A The population of organisms in the shallow water of lakes is high. Fewer types of organisms live in the deeper water.
B Ponds contain more vegetation than lakes contain.

Figure 17
Life in the Florida Everglades was threatened due to pollution, drought, and draining of the water. Conservation efforts are being made in an attempt to preserve this ecosystem.

Water Pollution Human activities can harm freshwater environments. Fertilizer-filled runoff from farms and lawns, as well as sewage dumped into the water, can lead to excessive growth of algae and plants in lakes and ponds. The growth and decay of these organisms reduces the oxygen level in the water, which makes it difficult for some organisms to survive. To prevent problems, sewage is treated before it is released. People also are being educated about problems associated with polluting lakes and ponds. Fines and penalties are issued to people caught polluting waterways. These controls have led to the recovery of many freshwater ecosystems.

Wetlands As the name suggests, **wetlands,** shown in **Figure 17,** are regions that are wet for all or most of a year. They are found in regions that lie between land masses and water. Other names for wetlands include swamps, bogs, and fens. Some people refer to wetlands as biological supermarkets. They are fertile ecosystems, but only plants that are adapted to water-logged soil survive there. Wetland animals include beavers, muskrats, alligators, and the endangered bog turtle. Many migratory bird populations use wetlands as breeding grounds.

✔ **Reading Check** *Where are wetlands found?*

Wetlands once were considered to be useless, disease-ridden places. Many were drained and destroyed to make roads, farmland, shopping centers, and housing developments. Only recently have people begun to understand the importance of wetlands. Products that come from wetlands, including fish, shellfish, cranberries, and plants, are valuable resources. Now many developers are restoring wetlands, and in most states access to land through wetlands is prohibited.

Chemistry
INTEGRATION

Osmosis is the movement of water molecules through a membrane from an area where there are more of them compared to other molecules present to an area where there are fewer of them compared to other molecules present. The blood of a freshwater fish is saltier than the lake water is. Explain in your Science Journal whether fish in a lake tend to absorb water or lose water through osmosis.

Saltwater Ecosystems

About 95 percent of the water on the surface of Earth contains high concentrations of various salts. The amount of dissolved salts in water is called salinity. Saltwater ecosystems include oceans, seas, a few inland lakes such as the Great Salt Lake in Utah, coastal inlets, and estuaries.

Math Skills Activity

Calculating Temperatures

Organisms that live around hydrothermal vents in the ocean deal with temperatures that range from 1.7°C to 371°C. You have probably seen temperatures measured in degrees Celsius (°C) and degrees Fahrenheit (°F). Which one are you familiar with? If you know the temperature in one system, you can convert it to the other.

Example Problem
You have a Fahrenheit thermometer and measure the water temperature of a pond at 59°F. What is that temperature in degrees Celsius?

Solution

1 *This is what you know:*
   water temperature in degrees Fahrenheit = 59°F

2 *This is what you want to find:*
   water temperature in degrees Celsius

3 *This is the equation you need to use:*
   (°C × 1.8) + 32 = °F

4 *Solve the equation for degrees Celsius and then substitute the known value:*
   °C = (°F − 32)/1.8
   °C = (59°F − 32)/1.8 = 15°C

Check your answer by substituting the Celsius temperature back into the original equation. Do you calculate the Fahrenheit temperature that was given?

Practice Problems

1. The thermometer outside your classroom reads 78°F. What is the temperature in degrees Celsius?

2. If lake water was 12°C in October and 23°C in May, what is the difference in degrees Fahrenheit?

For more help, refer to the Math Skill Handbook.

Open Oceans Life abounds in the open ocean. Scientists divide the ocean into different life zones, based on the depth to which sunlight penetrates the water. The lighted zone of the ocean is the upper 200 m or so. It is the home of the plankton that make up the foundation of the food chain in the open ocean. Below about 200 m is the dark zone of the ocean. Animals living in this region feed on material that floats down from the lighted zone, or they feed on each other. A few organisms are able to produce their own food.

Coral Reefs One of the most diverse ecosystems in the world is the **coral reef.** Coral reefs are formed over long periods of time from the calcium carbonate shells secreted by animals called corals. When corals die, their shells remain. Over time, the shell deposits form reefs such as the Great Barrier Reef off the coast of Australia, shown in **Figure 18.**

Reefs do not adapt well to long-term stress. Runoff from fields, sewage, and increased sedimentation from cleared land harm reef ecosystems. Organizations like the Environmental Protection Agency have developed management plans to protect the diversity of coral reefs. These plans treat a coral reef as a system that includes all the areas that surround the reef. Keeping the areas around reefs healthy will result in a healthy environment for the coral reef ecosystem.

Research Visit the Glencoe Science Web site at **science.glencoe.com** for more information about coral reefs. Communicate to your class what you learn.

Figure 18
A The lighter areas around this island are part of the Great Barrier Reef. It comprises about 3,000 reefs and about 900 islands. **B** Reefs contain colorful fish and a large variety of other organisms.

Figure 19
As the tide recedes, small pools of seawater are left behind. These pools contain a variety of organisms such as sea stars and periwinkles.

Earth Science INTEGRATION

Seashores All of Earth's landmasses are bordered by ocean water. The shallow waters along the world's coastlines contain a variety of saltwater ecosystems, all of which are influenced by the tides and by the action of waves. The gravitational pull of the Sun and Moon on Earth causes the tides to rise and fall each day. The height of the tides varies according to the phases of the Moon, the season, and the slope of the shoreline. The **intertidal zone** is the portion of the shoreline that is covered with water at high tide and exposed to the air during low tide. Organisms that live in the intertidal zone, such as those in **Figure 19,** must be adapted to dramatic changes in temperature, moisture, and salinity and must be able to withstand the force of wave action.

Estuaries Almost every river on Earth eventually flows into an ocean. The area where a river meets an ocean and contains a mixture of freshwater and salt water is called an **estuary** (ES chuh wer ee). Other names for estuaries include bays, lagoons, harbors, inlets, and sounds. They are located near coastlines and border the land. Salinity in estuaries changes with the amount of freshwater brought in by rivers and streams, and with the amount of salt water pushed inland by the ocean tides.

Estuaries, shown in **Figure 20,** are extremely fertile, productive environments because freshwater streams bring in tons of nutrients washed from inland soils. Therefore, nutrient levels in estuaries are higher than in freshwater ecosystems or other saltwater ecosystems.

Estuary Life Organisms found in estuaries include many species of algae, salt-tolerant grasses, shrimp, crabs, clams, oysters, snails, worms, and fish. Estuaries also serve as important nurseries for many species of ocean fish. They provide much of the seafood consumed by humans.

 Reading Check *What types of organisms live in estuaries?*

Section 3 Assessment

1. What are the similarities and differences between a lake and a stream?

2. Compare and contrast the dark zone of the ocean with the forest floor of a tropical rain forest. What living or nonliving factors affect these areas?

3. Why do you find few plants at the bottom of deep lakes?

4. What adaptations are necessary for organisms that live in the intertidal zone?

5. **Think Critically** Why do few plants grow in a swift-flowing mountain stream?

Skill Builder Activities

6. **Concept Mapping** Make an events chain concept map of these ecosystems from least salty to most salty: *lake, open ocean, estuary,* and *stream.* **For more help, refer to the** Science Skill Handbook.

7. **Communicating** Wetlands trap and slowly release rain, snow, and groundwater. Describe in your Science Journal what might happen to a town located on a floodplain if nearby wetlands are destroyed. **For more help, refer to the** Science Skill Handbook.

Activity *Use the Internet*

Exploring Wetlands

Wetlands, such as the one shown below, are an important part of the environment. These fertile ecosystems support unique plants and animals that can survive only in wetland conditions. The more you understand the importance of wetlands, the more you can do to preserve and protect them.

Recognize the Problem

Why are wetlands an important part of the ecosystem?

Form a Hypothesis

Why do wetlands need to be protected? What laws are in place to protect wetlands? Form a hypothesis about why wetlands should be protected.

Goals
- ■ **Identify** wetland regions in Texas and other parts of the United States.
- ■ **Describe** the significance of the wetland ecosystem.
- ■ **Identify** plant and animal species native to a wetland region.
- ■ **Identify** strategies for supporting the preservation of wetlands.

Data Source

SCIENCE*Online* Visit the Glencoe Science Web site at **science.glencoe.com** to get more information about wetland environments and for data collected by other students.

Test Your Hypothesis

Plan

1. Determine where some major wetlands are located in the United States.

2. Select one wetland area to study in depth. Where is it located? Is it classified as a marsh, bog, or something else?

3. What role does this ecosystem play in the overall ecology of the area?

4. Find out about the plants and animals that live in the wetland environment you are researching.

5. In most cases, there are federal laws to protect the environment. Additional state and local laws are specific to the area. Investigate what laws protect the wetland you are studying.

Do

1. Make sure your teacher approves your plan before you start.

2. Perform the investigation.

3. Go to the Glencoe Science Web site at **science.glencoe.com** to post your data.

Analyze Your Data

1. **Describe** the wetland area you have researched. What region of the United States is it located in? What other ecological factors are found in that region?

2. What laws protect the wetland you are investigating? How long have the laws been in place?

3. What plants and animals are native to the wetland area you are researching? Are those plants and animals found in other parts of the region or the United States? What adaptations do the plants and animals have that help them survive in a wetland environment?

Draw Conclusions

1. Are all wetlands the same?

2. What is the ecological significance of the wetland area that you studied for that region of the country?

3. Why should wetland environments be protected?

4. What can people do to support the continued preservation of wetland environments in the United States?

Communicating Your Data

SCIENCE Online Find this *Use the Internet* activity on the Glencoe Science Web site at **science. glencoe.com** and **post** your data in the table provided. **Review** other students' data to learn about other wetland environments in the United States.

TIME
SCIENCE AND
Society
SCIENCE
ISSUES
THAT AFFECT
YOU!

Helping Nature

Pebbles were added to the Corrales wetlands to help with drainage.

When you wash your hands or flush the toilet, you probably don't think much about where the wastewater goes. In most places, it eventually ends up being processed in a traditional sewage-treatment facility. But a handful of places are experimenting with a new method that processes wastewater by creating wetlands. Wetlands, such as swamps or marshes, are home to filtering plants, such as cattails, and sewage-eating bacteria.

In 1996, school officials at the Corrales Elementary School in the city of Albuquerque, New Mexico, faced a big problem. The old wastewater-treatment system had failed.

Replacing it was going to cost a lot of money. The school officials came up with an alternative plan. Instead of constructing a new sewage-treatment plant, they decided to create a natural wetlands system. The wetlands system could do the job less expensively, while protecting the environment. Today, this wetlands efficiently converts polluted water into cleaner water that's good for the environment. U.S. government officials are closely watching this alternative sewage-treatment system to see how successful it is. So far, so good! And if it continues to work well, other communities may start creating wetlands to filter their sewage.

Creating Wetlands to Purify Wastewater

Help Itself

Water irises thrived in the wetlands, less than a year after planting.

Students enjoy pure water from the Corrales wetlands after it is filtered.

How Wetlands Work

Wetlands filter water through the actions of the plants and microorganisms that live there. When plants absorb water into their roots, some, such as cattails, also take up pollutants. The plants convert the pollutants to forms that are not dangerous. At the same time, bacteria and other microorganisms are filtering water as they feed. Water does not move quickly through wetlands, so the organisms have plenty of time to do their work. The wetlands built at Corrales Elementary is called a "constructed" wetlands.

That is a wetlands built by people to filter small amounts of pollutants. In many places, constructed wetlands are better at cleaning wastewater than sewers or septic systems.

From Sewage to Flowers

Nancy Griego, a wetlands plant expert, planted a variety of colorful plants that bloom at different times of the year. A local official said, "School classes are taking field trips to the wetlands to learn what plants do as part of the environment. It's become a real educational asset."

CONNECTIONS Visit and Observe Visit a wetlands. Create a field journal of your observations. Sketch pictures of the plants and animals you see. **BONUS:** Use a field guide to help identify the wildlife you see. If you don't live near a wetlands, use resources to research wetlands environments.

SCIENCE Online
For more information, visit
science.glencoe.com

Reviewing Main Ideas

Section 1 How Ecosystems Change

1. Ecological succession is the gradual change from one community to another.

2. Primary succession begins in a place where no soil exists. It begins with organisms that are called pioneer species.

3. Secondary succession begins in a place that has soil and was once the home of living organisms. *What type of succession is occurring here?*

4. A community that has reached a stable stage of ecological succession is called a climax community.

Section 2 Biomes

1. Temperature and precipitation help determine the climate of a region.

2. Large geographic areas with similar climax communities are called biomes. *What type of biome is shown in the photo to the right?*

3. Earth's land biomes include tundra, taiga, temperate deciduous forest, temperate rain forest, tropical rain forest, grassland, and desert.

Section 3 Aquatic Ecosystems

1. Freshwater ecosystems include flowing water, such as streams and rivers, and standing water such as lakes, ponds, and wetlands.

2. Wetlands are land areas that are covered with water most of the year. They are found in regions that lie between land masses and water.

3. Saltwater ecosystems include estuaries, seashores, coral reefs, a few inland lakes, and the deep ocean. *Why are coral reefs, like this one, rich in biodiversity?*

4. Estuaries are important transitional zones between freshwater and saltwater environments. They are extremely fertile, and provide much of the seafood consumed by humans.

FOLDABLES Reading & Study Skills

After You Read

To help review succession, use the Main Ideas Study Fold that you made at the beginning of this chapter. Which type of succession starts with bare rock?

Visualizing Main Ideas

Fill in the circles of the concept map showing land biomes.

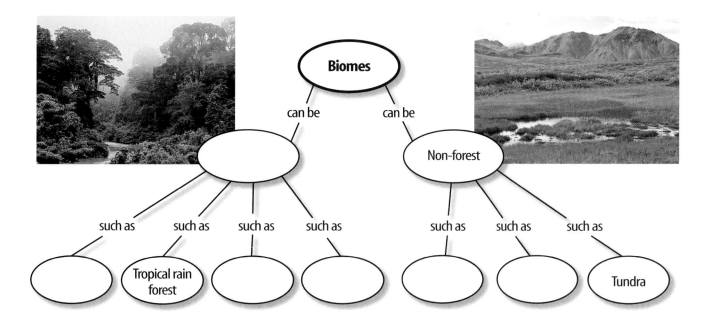

Vocabulary Review

Vocabulary Words

a. biome
b. climax community
c. coral reef
d. desert
e. estuary
f. grassland
g. intertidal zone
h. pioneer species
i. succession
j. taiga
k. temperate deciduous forest
l. temperate rain forest
m. tropical rain forest
n. tundra
o. wetland

Study Tip

Pay attention to the chapter's illustrations. Try to figure out exactly what main point the picture is trying to stress.

Using Vocabulary

Each of the following sentences is false. Make the sentence true by replacing the underlined word with the correct vocabulary word.

1. <u>Biome</u> refers to the normal changes in the types of species that live in communities.

2. A <u>pioneer species</u> is a group of organisms found in a stable stage of succession.

3. Deciduous trees are dominant in the <u>temperate rain forest</u>.

4. The average temperature in <u>tropical rain forests</u> is between 9°C and 12°C.

5. <u>Taigas</u> are the most biologically diverse biomes in the world.

6. An <u>intertidal zone</u> is an area where freshwater meets the ocean.

Chapter 3 Assessment

Checking Concepts

Choose the word or phrase that best answers the question.

1. What are tundra and desert examples of?
   A) ecosystems   **C)** habitats
   B) biomes   **D)** communities

2. What is a hot, dry biome called?
   A) desert   **C)** coral reef
   B) tundra   **D)** grassland

3. Where would organisms that are adapted to live in freshwater and salt water be found?
   A) lake   **C)** open ocean
   B) estuary   **D)** intertidal zone

4. Which biome contains the largest number of species?
   A) taiga
   B) temperate deciduous forest
   C) tropical rain forest
   D) grassland

5. Which biome contains mostly frozen soil called permafrost?
   A) taiga
   B) temperate rain forest
   C) tundra
   D) temperate deciduous forest

6. A new island is formed from a volcanic eruption. Which species probably would be the first to grow and survive?
   A) oak trees   **C)** grasses
   B) lichens   **D)** ferns

7. What would the changes in communities that take place on a recently formed volcanic island best be described as?
   A) primary succession
   B) secondary succession
   C) tertiary succession
   D) magma

8. What is the stable stage of succession?
   A) pioneer species   **C)** limiting factor
   B) climax community   **D)** permafrost

9. Which of the following aquatic ecosystems would have the lowest salinity?
   A) estuaries   **C)** coral reefs
   B) rivers   **D)** open oceans

10. Which ecosystem would experience the most drastic changes in a 24-hour period?
    A) open ocean   **C)** lake
    B) intertidal zone   **D)** coral reef

Thinking Critically

11. In most cases, would a soil sample from a temperate deciduous forest be more or less nutrient-rich than a soil sample from a tropical rain forest? Explain.

12. Why do many tropical rain forest plants make good houseplants?

13. Some plant seeds need fire in order to germinate. Explain how this gives these plants an advantage in secondary succession.

14. A grassy meadow borders a beech-maple forest. Is one of these ecosystems undergoing succession? Why?

15. Why are tundra plants usually small?

Developing Skills

16. **Concept Mapping** Make a network-tree concept map for water environments. Include these terms: *saltwater ecosystems, freshwater ecosystems, intertidal zone, lighted zone, dark zone, lake, pond, coral reef, river, stream, flowing water,* and *standing water.*

17. **Comparing and Contrasting** Compare and contrast the adaptations of organisms living in the tundra with those living in the taiga.

18. **Forming Hypotheses** Make a hypothesis about what would happen to succession in a pond if the pond owner removed all the cattails and reeds from around the pond edges every summer.

19. **Recognizing Cause and Effect** Devastating fires, like the one in Yellowstone National Park in 1988, cause many changes to the land. Determine the effect of a fire on an area that has reached its climax community.

20. **Making and Using Graphs** Make a graph of the amount of rainfall per year in each biome. Which type of graph is best suited to display the data?

Rainfall Amounts	
Biome	**Rainfall/Year (cm)**
Tundra	25
Taiga	50
Temperate rain forest	200
Tropical rain forest	400
Grassland	35
Temperate deciduous forest	150
Desert	25

Performance Assessment

21. **Oral Presentation** Research a biome that was not covered in this chapter. Find out about the climate, where it is located, and which organisms live there. Present this information to your class.

TECHNOLOGY

Go to the Glencoe Science Web site at **science.glencoe.com** or use the **Glencoe Science CD-ROM** for additional chapter assessment.

THE PRINCETON REVIEW **Test Practice**

Each biome around the world contains many different groups of organisms that are unique to that biome. Plants and animals from two different biomes have been drawn in separate boxes below.

A   **B**

Study the pictures above and answer the following questions.

1. The animals in Box B are different from the animals in Box A because the animals in Box B have adaptations that _____.
   A) allow them to survive during heavy rainfall
   B) allow them to reproduce asexually and sexually
   C) allow them to conserve water
   D) allow them to maintain a constant internal temperature

2. Which of the following organisms would be most likely to complete the group shown in Box B?
   F) palm tree
   G) salamander
   H) fish
   J) snake

Conserving Resources

I t's sweltering and you haven't felt a breeze in hours. You lift the hose over your head and turn the faucet handle. Instead of clear, cool water, greenish-yellow sludge flows out. Yuck! Clean water isn't the only resource that is commonly taken for granted. In this chapter, you'll learn the difference between renewable and nonrenewable resources. You'll also read about the various energy sources available on Earth and how pollution can affect your life.

What do you think?

Science Journal Look at the picture below with a classmate. Discuss what this might be. Here's a hint: *It's enriching.* Write your answer or best guess in your Science Journal.

Land, earth, dirt, soil—the ground beneath your feet is called by many names. You walk on it. People build roads and buildings on it. Plants grow in the top, nutrient-rich layer, called topsoil. Plants help keep topsoil in place by protecting it from wind and rain. What happens when topsoil is left unprotected?

Model topsoil loss

1. Use a mixture of moist sand and potting soil to create a miniature landscape in a plastic basin or aluminum-foil baking pan. Form hills and valleys in your landscape.

2. Use clumps of moss to cover areas of your landscape. Leave some sloping portions without plant cover.

3. Simulate a rainstorm over your landscape by spraying water on it from a spray bottle or by pouring a slow stream of water on it from a beaker.

Observe

In your Science Journal record your observations and describe what happened to the land that was not protected by plant cover.

Before You Read

FOLDABLES
Reading & Study
Skills

Making a Concept Map Study Fold Make the following Foldable to help you organize information and diagram ideas about renewable and nonrenewable resources.

1. Place a sheet of paper in front of you so the short side is at the top. Fold the bottom of the paper to the top, stopping about four centimeters from the top.

2. Draw an oval above the fold. Write *Natural Resources* inside the oval.

3. Fold the paper in half from the left side to the right side and then unfold. Through the top thickness of the paper, cut along the fold line to form two tabs.

4. Label the tabs *Renewable* and *Nonrenewable* and draw an oval around each word. Draw arrows from the large oval to the smaller ovals.

5. Before you read the chapter, list examples of each type of natural resource you already know on the front of the tabs. As you read the chapter, add to your lists.

SECTION
1 Resources

As You Read

What You'll Learn

- **Compare** renewable and nonrenewable resources.
- **List** uses of fossil fuels.
- **Identify** alternatives to fossil fuel use.

Vocabulary

natural resource
renewable resource
nonrenewable resource
petroleum
fossil fuel
hydroelectric power
nuclear energy
geothermal energy

Why It's Important

Wise use of natural resources is important for the health of all life on Earth.

Natural Resources

An earthworm burrowing in moist soil eats decaying plant material. A robin catches the worm and flies to a tree. The leaves of the tree use sunlight during photosynthesis. Leaves fall to the ground, decay, and perhaps become an earthworm's meal. What do these living things have in common? They rely on Earth's **natural resources**—the parts of the environment that are useful or necessary for the survival of living organisms.

What kinds of natural resources do you use? Like other organisms, you need food, air, and water. You also use resources that are needed to make everything from clothes to cars. Natural resources supply energy for automobiles and power plants. Although some natural resources are plentiful, others are not.

Renewable Resources The Sun, an inexhaustible resource, provides a constant supply of heat and light. Rain fills lakes and streams with water. When plants carry out photosynthesis, they add oxygen to the air. Sunlight, water, air, and the crops shown in **Figure 1** are examples of renewable resources. A **renewable resource** is any natural resource that is recycled or replaced constantly by nature.

Figure 1

Cotton and wood are renewable resources. **A** Cotton cloth is used for rugs, curtains, and clothing. A new crop of cotton can be grown every year.
B Wood is used for furniture, building materials, and paper. It will take 20 years for these young trees to grow large enough to harvest.

Supply and Demand Even though renewable resources are recycled or replaced, they are sometimes in short supply. Rain and melted snow replace the water in streams, lakes, and reservoirs. Sometimes, there may not be enough rain or snowmelt to meet all the needs of people, plants, and animals. In some parts of the world, especially desert regions, water and other resources usually are scarce. Other resources can be used instead, as shown in **Figure 2.**

Nonrenewable Resources Natural resources that are used up more quickly than they can be replaced by natural processes are **nonrenewable resources.** Earth's supply of nonrenewable resources is limited. You use nonrenewable resources when you take home groceries in a plastic bag, paint a wall, or travel by car. Plastics, paints, and gasoline are made from an important nonrenewable resource called petroleum, or oil. **Petroleum** is formed mostly from the remains of microscopic marine organisms buried in Earth's crust. It is nonrenewable because it takes hundreds of millions of years for it to form.

✔ **Reading Check** *What are nonrenewable resources?*

Minerals and metals found in Earth's crust are nonrenewable resources. Petroleum is a mineral. So are diamonds and the graphite in pencil lead. The aluminum used to make soft-drink cans is a metal. Iron, copper, tin, gold, silver, tungsten, and uranium also are metals. Many manufactured items, like the car shown in **Figure 3,** are made from nonrenewable resources.

Figure 2
In parts of Africa, firewood has become scarce. People in this village now use solar energy instead of wood for cooking.

Figure 3
Iron, a nonrenewable resource, is the main ingredient in steel. Steel is used to make cars, trucks, appliances, buildings, bridges, and even tires. *What other nonrenewable resources are used to build a car?*

Procedure 🔊 📬

1. Place a **chocolate chip cookie** on a **paper plate**. Pretend the chips are mineral deposits and the rest of the cookie is Earth's crust.

2. Use a **toothpick** to locate and dig up the mineral deposits. Try to disturb the land as little as possible.

3. When mining is completed, try to restore the land to its original condition.

Analysis

1. How well were you able to restore the land?

2. Compare the difficulty of digging for mineral deposits found close to the surface with digging for those found deep in Earth's crust.

3. Describe environmental changes that might result from a mining operation.

Fossil Fuels

Coal, oil, and natural gas are nonrenewable resources that supply energy. Most of the energy you use comes from these fossil fuels, as the graph in **Figure 4** shows. **Fossil fuels** are fuels formed in Earth's crust over hundreds of millions of years. Cars, buses, trains, and airplanes are powered by gasoline, diesel fuel, and jet fuel, which are made from oil. Coal is used in many power plants to produce electricity. Natural gas is used in manufacturing, for heating and cooking, and sometimes as a vehicle fuel.

Fossil Fuel Conservation Billions of people all over the world use fossil fuels every day. Because fossil fuels are non-renewable, Earth's supply of them is limited. In the future, they may become more expensive and difficult to obtain. Also, the use of fossil fuels can lead to environmental problems. For example, mining coal can require stripping away thick layers of soil and rock, as shown in **Figure 4,** which destroys ecosystems. Another problem is that fossil fuels must be burned to release the energy stored in them. The burning of fossil fuels produces waste gases that cause air pollution, including smog and acid rain. For these reasons, many people suggest reducing the use of fossil fuels and finding other sources of energy.

You can use simple conservation measures to help reduce fossil fuel use. Switch off the light when you leave a room and turn off the television when you're not watching it. These actions reduce your use of electricity, which often is produced in power plants that burn fossil fuels. Hundreds of millions of automobiles are in use in the United States. Riding in a car pool or taking public transportation uses fewer liters of gasoline than driving alone in a car. Walking or riding a bicycle uses even less fossil fuel. Reducing fossil fuel use has an added benefit—the less you use, the more money you save.

Figure 4

Coal is a fossil fuel. It often is obtained by strip mining, which removes all the soil above the coal deposit. The soil is replaced, but it takes many years for the ecosystem to recover. The graph shows that 84 percent of the energy used in the United States in 1999 came from fossil fuels.

Figure 5

Most power plants use turbine generators to produce electricity. In fossil fuel plants, burning fuel boils water and produces steam that turns the turbine.

Fast-moving steam, water, or wind rushes across the turbine blades. This flow of energy causes the turbine blades to turn.

The turbine blades are attached to a shaft. When the blades turn, so does the shaft.

Electricity flows from the coil into electrical wires.

Magnet

Generator

The turning shaft is connected to an electric generator. A simple generator is a coil of wire that spins inside the field of a magnet. The turbine shaft spins the coil. The spinning coil generates electricity.

Turbine

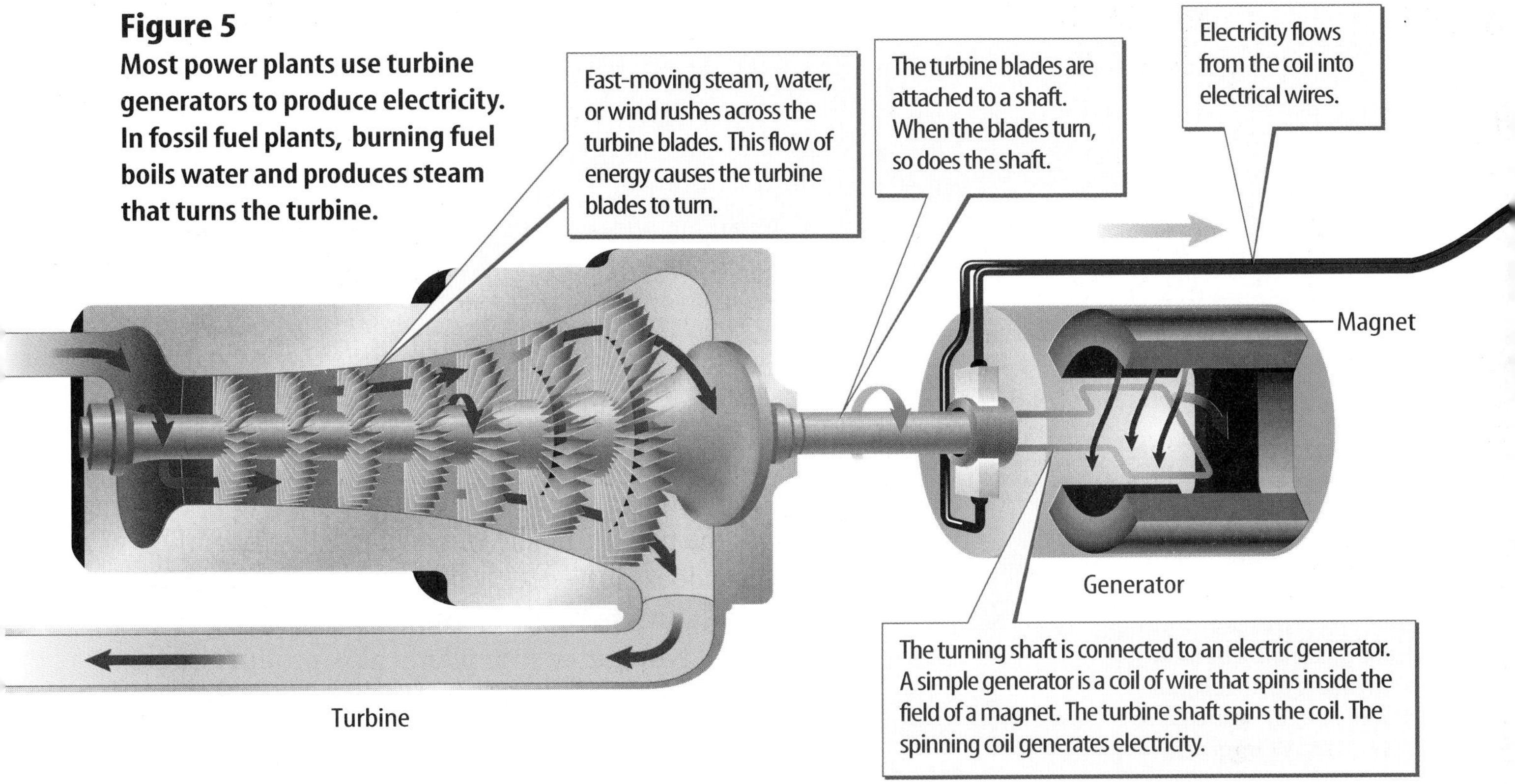

Alternatives to Fossil Fuels

Another approach to reducing fossil fuel use is to develop other sources of energy. Much of the electricity used today comes from power plants that burn fossil fuels. As **Figure 5** shows, electricity is generated when a rotating turbine turns a coil of wires in the magnetic field of an electric generator. Fossil-fuel power plants boil water to produce steam that turns the turbine. Alternative energy sources, including water, wind, and atomic energy can be used instead of fossil fuels to turn turbines. Also, solar cells can produce electricity using only sunlight, with no turbines at all. Some of these alternative energy sources—particularly wind and solar energy—are so plentiful they could be considered inexhaustible resources.

Water Power Water is a renewable energy source that can be used to generate electricity. **Hydroelectric power** is electricity that is produced when the energy of falling water is used to turn the turbines of an electric generator. Hydroelectric power does not contribute to air pollution because no fuel is burned. However, it does present environmental concerns. Building a hydroelectric plant usually involves constructing a dam across a river. The dam raises the water level high enough to produce the energy required for electricity generation. Many acres behind the dam are flooded, destroying land habitats and changing part of the river into a lake.

Physics
INTEGRATION

Potential energy is stored energy. Kinetic energy is energy in motion, like a car moving along a street. In your Science Journal, explain why water stored behind a dam has potential energy. Describe how this potential energy becomes kinetic energy.

Wind Power Wind power is another renewable energy source that can be used for electricity production. Wind turns the blades of a turbine, which powers an electric generator. When winds blow at least 32 km/h, energy is produced. Wind power does not cause air pollution, but electricity can be produced only when the wind is blowing. So far, wind power accounts for only a small percentage of the electricity used worldwide.

Nuclear Power Another alternative to fossil fuels makes use of the huge amounts of energy in the nuclei of atoms, as shown in **Figure 6. Nuclear energy** is released when billions of atomic nuclei from uranium, a radioactive element, are split apart in a nuclear fission reaction. This energy is used to produce steam that rotates the turbine blades of an electric generator.

Nuclear power does not contribute to air pollution. However, uranium is a nonrenewable resource, and mining it can disrupt ecosystems. Nuclear power plants also produce radioactive wastes that can seriously harm living organisms. Some of these wastes remain radioactive for thousands of years, and their safe disposal is a problem that has not yet been solved. Accidents are also a danger.

Figure 6
Nuclear power plants are designed to withstand the high energy produced by nuclear reactions.

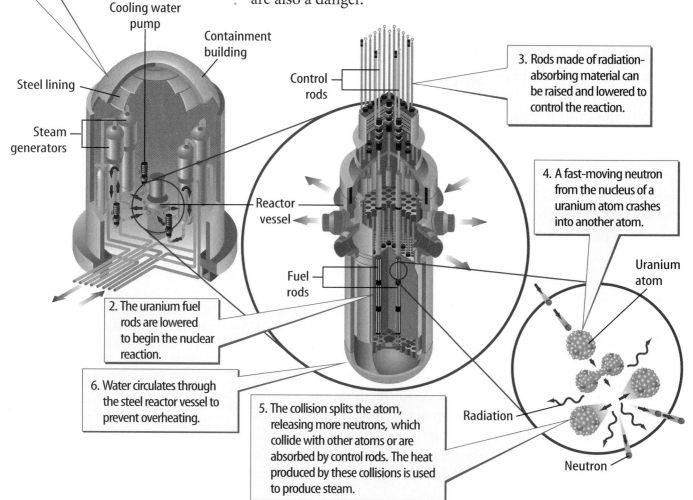

1. The containment building is made of concrete lined with steel. The reactor vessel and steam generators are housed inside.

Cooling water pump

Containment building

Steel lining

Steam generators

Control rods

3. Rods made of radiation-absorbing material can be raised and lowered to control the reaction.

4. A fast-moving neutron from the nucleus of a uranium atom crashes into another atom.

Reactor vessel

Uranium atom

Fuel rods

2. The uranium fuel rods are lowered to begin the nuclear reaction.

6. Water circulates through the steel reactor vessel to prevent overheating.

5. The collision splits the atom, releasing more neutrons, which collide with other atoms or are absorbed by control rods. The heat produced by these collisions is used to produce steam.

Radiation

Neutron

Geothermal Energy The hot, molten rock that lies deep beneath Earth's surface is also a source of energy. You see the effects of this energy when lava and hot gases escape from an erupting volcano or when hot water spews from a geyser. The heat energy contained in Earth's crust is called **geothermal energy.** Most geothermal power plants use this energy to produce steam to generate electricity.

Geothermal energy for power plants is available only where natural geysers or volcanoes are found. A geothermal power plant in California uses steam produced by geysers. The island nation of Iceland was formed by volcanoes, and geothermal energy is plentiful there. Geothermal power plants supply heat and electricity to about 90 percent of the homes in Iceland. Outdoor swimming areas also are heated with geothermal energy, as shown in **Figure 7.**

Solar Energy The most inexhaustible source of energy for all life on Earth is the Sun. Solar energy is an alternative to fossil fuels. One use of solar energy is in solar-heated buildings. During winter in the northern hemisphere, the parts of a building that face south receive the most sunlight. Large windows placed on the south side of a building help heat it by allowing warm sunshine into the building during the day. Floors and walls of most solar-heated buildings are made of materials that absorb heat during the day. During the night, the stored heat is released slowly, keeping the building warm. **Figure 8** shows how solar energy can be used.

Figure 7
In Iceland, a geothermal power plant pumps hot water out of the ground to heat buildings and generate electricity. Leftover hot water goes into this lake, making it warm enough for swimming even when the ground is covered with snow.

Figure 8
The Zion National Park Visitor Center in Utah is a solar-heated building designed to save energy. The roof holds solar panels that are used to generate electricity. High windows can be opened to circulate air and help cool the building on hot days. The overhanging roof shades the windows during summer.

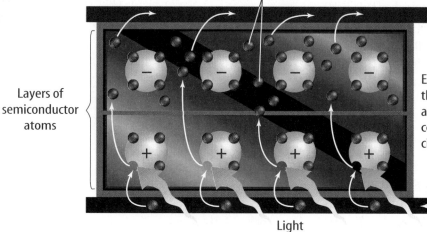

Free electrons

Layers of semiconductor atoms

Electric current flows through the calculator and back to the PV cell to form a complete circuit.

－

＋

Light

Figure 9
Light energy from the Sun travels in tiny packets of energy called photons. Photons crash into the atoms of PV cells, knocking electrons loose. These electrons create an electric current.

Solar Cells Do you know how a solar-powered calculator works? How do spacecraft use sunlight to generate electricity? These devices use photovoltaic (foht oh vohl TAY ihk) cells to turn sunlight into electric current, as shown in **Figure 9.** Photovoltaic (PV) cells are small and easy to use. However, they produce electricity only in sunlight, so batteries are needed to store electricity for use at night or on cloudy days. Also, PV cells presently are too expensive to use for generating large amounts of electricity. Improvements in this technology continue to be made, and prices probably will go down in the future. As **Figure 10** shows, solar buildings and PV cells are just two of the many ways solar energy can be used to replace fossil fuels.

Section 1 Assessment

1. What are natural resources?
2. Compare and contrast renewable and nonrenewable resources. Give five examples of each.
3. Name five energy sources that provide alternatives to fossil fuels.
4. Describe two ways solar energy can be used to reduce fossil fuel use.
5. **Think Critically** Explain why the water that is used to cool the reactor vessel of a nuclear power plant is kept separate from the water that is heated to produce steam for the turbine generators.

Skill Builder Activities

6. **Concept Mapping** Draw a concept map showing how the following terms are related: *renewable resources, nonrenewable resources, fossil fuels, natural gas, coal, oil, solar energy, PV cells, solar cookers, nuclear energy,* and *geothermal energy.* **For more help, refer to the Science Skill Handbook.**

7. **Solving One-Step Equations** Most cars in the U.S. are driven about 10,000 miles each year. If a car can travel 30 miles on one gallon of gasoline, how many gallons will it use in a year? **For more help, refer to the Math Skill Handbook.**

Figure 10

Sunlight is a renewable energy source that provides an alternative to fossil fuels. Solar technologies use the Sun's energy in many ways—from heating to electricity generation.

▼ **ELECTRICITY** Photovoltaic (PV) cells turn sunlight into electric current. They are commonly used to power small devices, such as calculators. Panels that combine many PV cells provide enough electricity for a home—or an orbiting satellite, such as the International Space Station, below.

▲ **POWER PLANTS** In California's Mojave Desert, an experimental solar power plant used hundreds of mirrors to focus sunlight on a water-filled tower. The steam produced by this system generates enough electricity to power 2,400 homes.

▼ **COOKING** In hot, sunny weather, a solar oven or panel cooker can be used to cook a pot of rice or heat water. The powerful solar cooker shown below reaches even higher temperatures. It is being used to fry food.

▼ **INDOOR HEATING** South-facing windows and heat-absorbing construction materials turn a room into a solar collector that can help heat an entire building, such as this Connecticut home.

▲ **WATER HEATING** Water is heated as it flows through small pipes in this roof-mounted solar heat collector. The hot water then flows into an insulated tank for storage.

What You'll Learn

- **Describe** types of air pollution.
- **Identify** causes of water pollution.
- **Explain** methods that can be used to prevent erosion.

Vocabulary
pollutant
acid precipitation
greenhouse effect
ozone depletion
erosion
hazardous waste

Why It's Important
By understanding the causes of pollution, you can help solve pollution problems.

Keeping the Environment Healthy

More than six billion people live on Earth. This large human population puts a strain on the environment, but each person can make a difference. You can help safeguard the environment by paying attention to how your use of natural resources affects air, land, and water.

Air Pollution

On a still, sunny day in almost any large city, you might see a dark haze in the air, like that in **Figure 11.** The haze comes from pollutants that form when wood or fuels are burned. A **pollutant** is a substance that contaminates the environment. Air pollutants include soot, smoke, ash, and gases such as carbon dioxide, carbon monoxide, nitrogen oxides, and sulfur oxides. Wherever cars, trucks, airplanes, factories, homes, or power plants are found, air pollution is likely. Air pollution also can be caused by volcanic eruptions, wind-blown dust and sand, forest fires, and the evaporation of paints and other chemicals.

Smog is a form of air pollution created when sunlight reacts with pollutants produced by burning fuels. It can irritate the eyes and make breathing difficult for people with asthma or other lung diseases. Smog can be reduced if people take buses or trains instead of driving or if they use vehicles, such as electric cars, that produce fewer pollutants than gasoline-powered vehicles.

Figure 11
The term *smog* was used for the first time in the early 1900s to describe the mixture of smoke and fog that often covers large cities in the industrial world.

Figure 12

A Compare these two photographs of the same statue. The photo on the left was taken before acid rain became a problem. The photo on the right shows acid rain damage.
B The pH scale indicates whether a solution is acidic or basic.

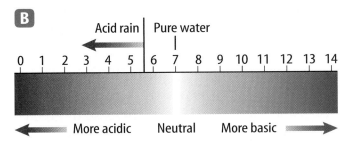

B

Acid rain | Pure water

0 1 2 3 4 5 | 6 7 8 9 10 11 12 13 14

← More acidic   Neutral   More basic →

Acid Precipitation

Chemistry **INTEGRATION**

Water vapor condenses on dust particles in the air to form droplets that combine to create clouds. Eventually, the droplets become large enough to fall to the ground as precipitation—mist, rain, snow, sleet, or hail. Air pollutants from the burning of fossil fuels can react with water in the atmosphere to form strong acids. Acidity is measured by a value called pH, as shown in **Figure 12. Acid precipitation** has a pH below 5.6.

Effects of Acid Rain Acid precipitation washes nutrients from the soil, which can lead to the death of trees and other plants. Runoff from acid rain that flows into a lake or pond can lower the pH of the water. If algae and microscopic organisms cannot survive in the acidic water, fish and other organisms that depend on them for food also die.

Preventing Acid Rain Sulfur from burning coal and nitrogen oxides from vehicle exhaust are the pollutants primarily responsible for acid rain. Using low-sulfur fuels, such as natural gas or low-sulfur coal, can help reduce acid precipitation. However, these fuels are less plentiful and more expensive than high-sulfur coal. Smokestacks that remove sulfur dioxide before it enters the atmosphere also help. Reducing automobile use and keeping car engines properly tuned can reduce acid rain caused by nitrogen oxide pollution. The use of electric cars, or hybrid-fuel cars that can run on electricity as well as gasoline, also could help.

Mini LAB

Measuring Acid Rain

Procedure

1. Collect **rainwater** by placing a clean **cup** outdoors. Do not collect rainwater that has been in contact with any object or organism.
2. Dip a piece of **pH indicator paper** into the sample.
3. Compare the color of the paper to the pH chart provided. Record the pH of the rainwater.
4. Use separate pieces of pH paper to test the pH of **tap water** and **distilled water**. Record these results.

Analysis

1. Is the rainwater acidic, basic, or neutral?
2. How does the pH of the rainwater compare with the pH of tap water? With the pH of distilled water?

Greenhouse Effect

Sunlight travels through the atmosphere to Earth's surface. Some of this sunlight normally is reflected back into space. The rest is trapped by certain atmospheric gases, as shown in **Figure 13.** This heat-trapping feature of the atmosphere is the **greenhouse effect.** Without it, temperatures on Earth probably would be too cold to support life.

Atmospheric gases that trap heat are called greenhouse gases. One of the most important greenhouse gases is carbon dioxide (CO_2). CO_2 is a normal part of the atmosphere. It is also a waste product that forms when fossil fuels are burned. Over the past century, more fossil fuels have been burned than ever before, which is increasing the percentage of CO_2 in the atmosphere. The atmosphere might be trapping more of the Sun's heat, making Earth warmer. A rise in Earth's average temperature, possibly caused by an increase in greenhouse gases, is known as global warming.

Global Warming

Temperature data collected from 1895 through 1995 indicate that Earth's average temperature increased about 1°C during that 100-year period. No one is certain whether this rise was caused by human activities or is a natural part of Earth's weather cycle. What kinds of changes might be caused by global warming? Changing rainfall patterns could alter ecosystems and affect the kinds of crops that can be grown in different parts of the world. The number of storms and hurricanes might increase. The polar ice caps might begin to melt, raising sea levels and flooding coastal areas. Warmer weather might allow tropical diseases, such as malaria, to become more widespread. Many people feel that the possibility of global warming is a good reason to reduce fossil fuel use.

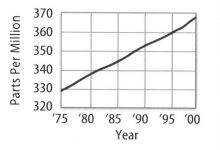

Carbon Dioxide Levels

Figure 13
The moment you step inside a greenhouse, you feel the results of the greenhouse effect. Heat trapped by the glass walls warms the air inside. In a similar way, atmospheric greenhouse gases trap heat close to Earth's surface.

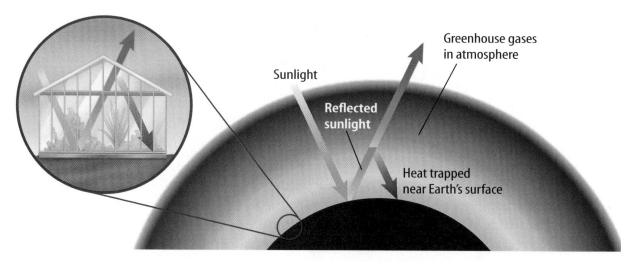

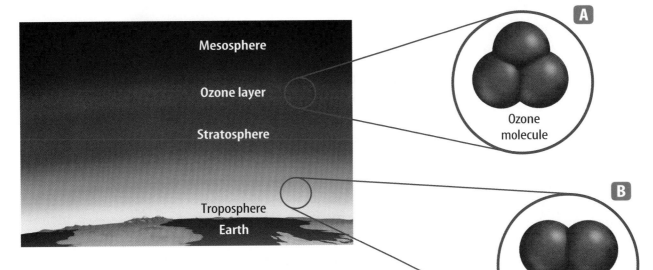

Figure 14
The atmosphere's ozone layer absorbs large amounts of UV radiation, preventing it from reaching Earth's surface.
A Ozone molecules are made of three oxygen atoms. They are formed in a chemical reaction between sunlight and oxygen.
B The oxygen you breathe has two oxygen atoms in each molecule.

Ozone Depletion

About 20 km above Earth's surface is a portion of the atmosphere known as the ozone (OH zohn) layer. Ozone is a form of oxygen, as shown in **Figure 14.** The ozone layer absorbs some of the Sun's harmful ultraviolet (UV) radiation. UV radiation can damage living cells.

Every year, the ozone layer temporarily becomes thinner over each polar region during its spring season. The thinning of the ozone layer is called **ozone depletion.** This problem is caused by certain pollutant gases, especially chlorofluorocarbons (klor oh FLOR oh kar bunz) (CFCs). CFCs are used in the cooling systems of refrigerators, freezers, and air conditioners. When CFCs leak into the air, they slowly rise into the atmosphere until they arrive at the ozone layer. CFCs react chemically with ozone, breaking apart the ozone molecules.

UV Radiation Because of ozone depletion, the amount of UV radiation that reaches Earth's surface could be increasing. UV radiation could be causing a rise in the number of skin cancer cases in humans. It also might be harming other organisms. The ozone layer is so important to the survival of life on Earth that world governments and industries have agreed to stop making and using CFCs.

Ozone that is high in the upper atmosphere protects life on Earth. Near Earth's surface though, it can be harmful. Ozone is produced when fossil fuels are burned. This ozone stays in the lower atmosphere, where it pollutes the air. Ozone damages the lungs and other sensitive tissues of animals and plants. For example, it can cause the needles of a Ponderosa pine to drop, harming growth.

✔ **Reading Check** *What is the difference between ozone in the upper atmosphere and ozone in the lower atmosphere?*

Indoor Air Pollution

Air pollution can occur indoors. Today's buildings are better insulated to conserve energy. However, better insulation reduces the flow of air into and out of a building, so air pollutants can build up indoors. For example, burning cigarettes release hazardous particles and gases into the air. Even non-smokers can suffer ill effects from secondhand cigarette smoke. As a result, smoking no longer is allowed in many public and private buildings. Paints, carpets, glues and adhesives, printers, and photocopy machines also give off dangerous gases, including formaldehyde. Like cigarette smoke, formaldehyde is a carcinogen, which means it can cause cancer.

Carbon Monoxide Carbon monoxide (CO) is a poisonous gas that is produced whenever charcoal, natural gas, kerosene, or other fuels are burned. CO poisoning can cause serious illness or death. Fuel-burning stoves and heaters must be designed to prevent CO from building up indoors. CO is colorless and odorless, so it is difficult to detect. Alarms that provide warning of a dangerous buildup of CO are being used in more and more homes.

Radon Radon is a naturally occurring, radioactive gas that is given off by some types of rock and soil, as shown in **Figure 15**. Radon has no color or odor. It can seep into basements and the lower floors of buildings. Radon exposure is the second leading cause of lung cancer in this country. A radon detector sounds an alarm when levels of the gas in indoor air become too high. If radon is present, increasing a building's ventilation can eliminate any damaging effects.

Figure 15

A The map shows the potential for radon exposure in different parts of the United States. Soils of the northern and northeastern portions of the country produce more radon gas than in most other areas. **B** A radon detection kit can tell the user if a dangerous level of radon is present.

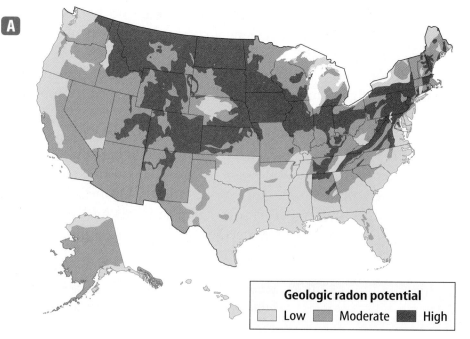

Geologic radon potential
Low   Moderate   High

A When rain falls on roads and parking lots, it can wash oil and grease onto the soil and into nearby streams.

B Rain can wash agricultural pesticides and fertilizers into lakes, streams, or oceans.

C Industrial wastes are sometimes released directly into surface waters.

Water Pollution

Pollutants enter water, too. Air pollutants can drift into water or be washed out of the sky by rain. Rain can wash land pollutants into waterways, as shown in **Figure 16.** Wastewater from factories and sewage-treatment plants often is released into waterways. In the United States and many other countries, laws require that wastewater be treated to remove pollutants before it is released. But, in many parts of the world, wastewater treatment is not always possible. Pollution also enters water when people dump litter or waste materials into rivers, lakes, and oceans.

Surface Water Some water pollutants poison fish and other wildlife, and can be harmful to people who swim in or drink the water. For example, chemical pesticides sprayed on farmland can wash into lakes and streams. These chemicals can harm the insects that fish, turtles, or frogs rely on for food. Shortages of food can lead to deaths among water-dwelling animals. Some pollutants, especially those containing mercury and other metals, can build up in the tissues of fish. Eating contaminated fish and shellfish can transfer these metals to people, birds, and other animals. In some areas, people are advised not to eat fish or shellfish taken from polluted waterways.

Algal blooms are another water pollution problem. Raw sewage and excess fertilizer contain large amounts of nitrogen. If they are washed into a lake or pond, they can cause the rapid growth of algae. When the algae die, they are decomposed by huge numbers of bacteria that use up much of the oxygen in the water. Fish and other organisms can die from a lack of oxygen in the water.

Figure 16
Pollution of surface waters can occur in several ways.

Figure 17
In 1996, the oil tanker *Sea Empress* spilled more than 72 million kg of oil into the sea along the coast of Wales. More than $40 million was spent on the cleanup effort, but thousands of ocean organisms were destroyed, including birds, fish, and shellfish.

Ocean Water Rivers and streams eventually flow into oceans, bringing their pollutants along. Also, polluted water can enter the ocean in coastal areas where factories, sewage-treatment plants, or shipping activities are located. Oil spills are a well-known ocean pollution problem. About 4 billion kg of oil are spilled into ocean waters every year. Much of that oil comes from ships that use ocean water to wash out their fuel tanks. Oil also can come from oil tanker wrecks, as shown in **Figure 17.**

Groundwater Pollution can affect water that seeps underground, as shown in **Figure 18.** Groundwater is water that collects between particles of soil and rock. It comes from precipitation and runoff that soaks into the soil. This water can flow slowly through permeable layers of rock called aquifers. If this water comes into contact with pollutants as it moves through the soil and into an aquifer, the aquifer could become polluted. Polluted groundwater is difficult—and sometimes impossible—to clean. In some parts of the country, chemicals leaking from underground storage tanks have created groundwater pollution problems.

Figure 18
Water from rainfall slowly filters through sand or soil until it is trapped in underground aquifers. Pollutants picked up by the water as it filters through the soil can contaminate water wells.

A Contour plowing reduces the downhill flow of water.

Soil Loss

Fertile topsoil is important to plant growth. New topsoil takes hundreds or thousands of years to form. The Explore Activity at the beginning of this chapter shows that rain washes away loose topsoil. Wind also blows it away. The movement of soil from one place to another is called **erosion** (ih ROH zhun). Eroded soil that washes into a river or stream can block sunlight and slow photosynthesis. It also can harm fish, clams, and other organisms. Erosion is a natural process, but human activities increase it. When a farmer plows a field or a forest is cut down, soil is left bare. Bare soil is more easily carried away by rain and wind. **Figure 19** shows some methods farmers use to reduce soil erosion.

Soil Pollution

Soil can become polluted when air pollutants drift to the ground or when water leaves pollutants behind as it flows through the soil. Soil also can be polluted when people toss litter on the ground or dispose of trash in landfills.

Solid Wastes What happens to the trash you throw out every week? What do people do with old refrigerators, TVs, and toys? Most of this solid waste is dumped in landfills. Most landfills are designed to seal out air and water. This helps prevent pollutants from seeping into surrounding soil, but it slows normal decay processes. Even food scraps and paper, which usually break down quickly, can last for decades in a landfill. In populated areas, landfills fill up quickly. Reducing the amount of trash people generate can reduce the need for new landfills.

Figure 19
The farming methods shown here help prevent soil erosion. *Why is soil erosion a concern for farmers?*

B On steep hillsides, flat areas called terraces reduce downhill flow.

C In strip cropping, cover crops are planted between rows to reduce wind erosion.

D In no-till farming, soil is never left bare.

Figure 20
Leftover paints, batteries, pesticides, drain cleaners, and medicines are hazardous wastes that should not be discarded in the trash. They should never be poured down a drain, onto the ground, or into a storm sewer. Most communities have collection facilities where people can dispose of hazardous materials like these.

Hazardous Wastes Waste materials that are harmful to human health or poisonous to living organisms are **hazardous wastes.** They include dangerous chemicals, such as pesticides, oil, and petroleum-based solvents used in industry. They also include radioactive wastes from nuclear power plants, from hospitals that use radioactive materials to treat disease, and from nuclear weapons production. Many household items also are considered hazardous like those shown in **Figure 20.** If these materials are dumped into landfills, they could seep into the soil, surface water, or groundwater over time. Hazardous wastes usually are handled separately from other types of trash. They are sealed in steel drums or treated in other ways to prevent them from polluting the environment.

✔ **Reading Check** *What are hazardous wastes?*

Section ② Assessment

1. List four ways in which air pollution affects the environment.

2. In what ways can an algal bloom affect pond life?

3. What methods can farmers use to prevent erosion?

4. Describe the possible causes and effects of ozone depletion.

5. **Think Critically** How could hazardous chemicals deposited in landfills eventually affect groundwater?

Skill Builder Activities

6. **Comparing and Contrasting** Compare and contrast the causes and effects of air and water pollution. **For more help, refer to the** Science Skill Handbook.

7. **Solving One-Step Equations** A pH of 4 is ten times more acidic than a pH of 5. A pH of 5 is ten times more acidic than a pH of 6. How many times more acidic is a solution with pH 4 than one with pH 6? **For more help, refer to the** Math Skill Handbook.

Activity

The Greenhouse Effect

You can create models of Earth with and without heat-reflecting greenhouse gases. Then, experiment with the models to observe the greenhouse effect.

What You'll Investigate
How does the greenhouse effect influence temperatures on Earth?

Materials
1-L clear, plastic, soft-drink bottle
  with top cut off and label removed (2)
thermometer (2)
*temperature probe
potting soil
masking tape
plastic wrap
rubber band
lamp with 100-W lightbulb
watch or clock with second hand
*Alternate materials

Goals
■ **Observe** the greenhouse effect.
■ **Describe** the effect that a heat source has on an environment.

Safety Precautions

Procedure
1. Copy the data table and use it to record your temperature measurements.
2. Put an equal volume of potting soil in the bottom of each container.
3. Use masking tape to attach a thermometer to the inside of each container. Place each thermometer at the same height above the soil.

Changes in Temperature		
Time (min)	Open Container Temperature (°C)	Closed Container Temperature (°C)
0		
2		
4		
6		

Shield each thermometer bulb by putting a double layer of masking tape over it.

4. Seal the top of one container with plastic wrap held in place with a rubber band.
5. Place the lamp with the exposed 100-W lightbulb between the two containers and exactly 1 cm away from each. Do not turn on the light.
6. Let the setup sit for 5 min, then record the temperature in each container.
7. Turn on the light. Record the temperature in each container every 2 min for 15 min to 20 min. Graph the results.

Conclude and Apply
1. **Compare and contrast** temperatures in each container at the end of the experiment.
2. What does the lightbulb represent in this experimental model? What does the plastic wrap represent?

Communicating
Your Data

Average the data obtained in the experiments conducted by all the groups in your class. Prepare a line graph of these data. **For more help,** refer to the Science Skill Handbook.

3 The Three Rs of Conservation

As You Read

What You'll Learn

- **Recognize** ways you can reduce your use of natural resources.
- **Explain** how you can reuse resources to promote conservation.
- **Describe** how many materials can be recycled.

Vocabulary
recycling

Why It's Important
Conservation preserves resources and reduces pollution.

Figure 21
Automobile tires are almost indestructible. They usually are disposed of but can have other useful purposes.

Conservation

A teacher travels to school in a car pool. In the school cafeteria, students place glass bottles and cans in separate containers from the rest of the garbage. Conservation efforts like these can help prevent shortages of natural resources, slow growth of landfills, reduce pollution levels, and save people money. Every time a new landfill is created, an ecosystem is disturbed. Reducing the need for landfills is a major benefit of conservation. The three Rs of conservation are reduce, reuse, and recycle.

Reduce

You contribute to conservation whenever you reduce your use of natural resources. You use less fossil fuel when you walk or ride a bicycle instead of taking the bus or riding in a car. If you buy a carton of milk, reduce your use of petroleum by telling the clerk you don't need a plastic bag to carry it in.

You also can avoid buying things you don't need. For example, most of the paper, plastic, and cardboard used to package items for display on store shelves is thrown away as soon as the product is brought home. You can look for products with less packaging or with packaging made from recycled materials. What are some other ways you can reduce your use of natural resources?

Reuse

Another way to help conserve natural resources is to use items more than once. Reusing an item means using it again without changing it or reprocessing it, as shown in **Figure 21.** Bring reusable canvas bags to the grocery store to carry home your purchases. Donate clothes you've outgrown to charity so that others can reuse them. Take reusable plates and utensils on picnics instead of disposable paper items.

Recycle

If you can't avoid using an item, and if you can't reuse it, the next best thing is to recycle it. **Recycling** is a form of reuse that requires changing or reprocessing an item or natural resource. If your city or town has a curbside recycling program, you already separate recyclables from the rest of your garbage. Materials that can be recycled include glass, metals, paper, plastics, and yard and kitchen waste.

✔ **Reading Check** *How is recycling different from reusing?*

Plastics Plastic is more difficult to recycle than other materials, mainly because several types of plastic are in use. A recycle code marked on every plastic container indicates the type of plastic it is made of. Plastic soft-drink bottles, like the one shown in **Figure 22,** are made of type 1 plastic and are the easiest to recycle. Most plastic bags are made of type 2 or type 4 plastic; they can be reused as well as recycled. Types 6 and 7 can't be recycled at all because they are made of a mixture of different plastics. Each type of plastic must be separated carefully before it is recycled because a single piece of a different type of plastic can ruin an entire batch.

Field **GUIDE**

What kinds of plastics can be recycled? To find out more about recycling, see the **Recyclable Plastics Field Guide** at the back of the book.

Figure 22
Many soft-drink bottles are made of PETE, which is the most common type of recyclable plastic. It can be melted down and spun into fibers to make carpets, paintbrushes, rope, and clothing.

SCIENCE *Online*

Research Visit the Glencoe Science Web site at **science.glencoe.com** and find out how to make your own recycled paper. In your Science Journal, describe how you might use the paper you make.

Metals The manufacturing industry has been recycling all kinds of metals, especially steel, for decades. At least 25 percent of the steel in cans, appliances, and automobiles is recycled steel. Up to 100 percent of the steel in plates and beams used to build skyscrapers is made from reprocessed steel. About one metric ton of recycled steel saves about 1.1 metric tons of iron ore and 0.5 metric ton of coal. Using recycled steel to make new steel products reduces energy use by 75 percent. Other metals, including iron, copper, aluminum, and lead also can be recycled.

You can conserve metals by recycling food cans, which are mostly steel, and aluminum cans. It takes less energy to make a can from recycled aluminum than from raw materials. Also, remember that recycled cans do not take up space in landfills.

Glass Glass bottles and jars can be sterilized and reused. They also can be melted and re-formed into new bottles, especially those made of clear glass. Most glass bottles already contain at least 25 percent recycled glass. Glass can be recycled again and again. It never needs to be thrown away. Recycling about one metric ton of glass saves more than one metric ton of mineral resources and cuts the energy used to make new glass by 25 percent or more.

Problem-Solving Activity

What items are you recycling at home?

Many people participate in community recycling programs. Recyclable items such as plastic, glass, newspapers, and metals may be picked up at the curbside or the resident may hire a licensed recycling handler to pick them up. What do you recycle in your home?

Identifying the Problem

The following bar graph shows the recycling rates in the U. S. of six types of household items for the years 1992, 1994, and 1995. How do your and your classmates' recycling rates compare with the recycling rates shown on the chart?

Solving the Problem

1. For one week, list each glass, plastic, and aluminum item you use. Note which items you throw away and which ones you recycle.

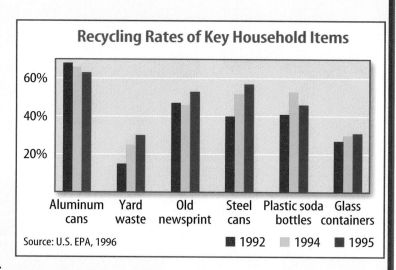

Calculate the percentage of glass, plastic, and aluminum you recycled. How do your percentages compare with those on the graph?

Paper Used paper is recycled into paper towels, insulation, newsprint, cardboard, and stationery. Ranchers and dairy farmers sometimes use shredded paper instead of straw for bedding in barns and stables. Used paper can be made into compost. Recycling about one metric ton of paper saves 17 trees, more than 26,000 L of water, close to 1,900 L of oil, and more than 4,000 kw of electric energy. You can do your part by recycling newspapers, notebook and printer paper, cardboard, and junk mail.

✔ Reading Check *What nonrenewable resource(s) do you conserve by recycling paper?*

Figure 23
Composting is a way of turning plant material you would otherwise throw away into rich garden soil. Dry leaves and weeds, grass clippings, vegetable trimmings, and nonmeat food scraps can be composted.

Compost Grass clippings, leaves, and fruit and vegetable scraps that are discarded in a landfill can remain there for decades without breaking down. The same items can be turned into soil-enriching compost in just a few weeks, as shown in **Figure 23.** Many communities distribute compost bins to encourage residents to recycle fruit and vegetable scraps and yard waste.

Buy Recycled People have become so good at recycling that recyclable materials are piling up, just waiting to be put to use. You can help by reading labels when you shop and choosing products that contain recycled materials. What other ways of recycling natural resources can you think of?

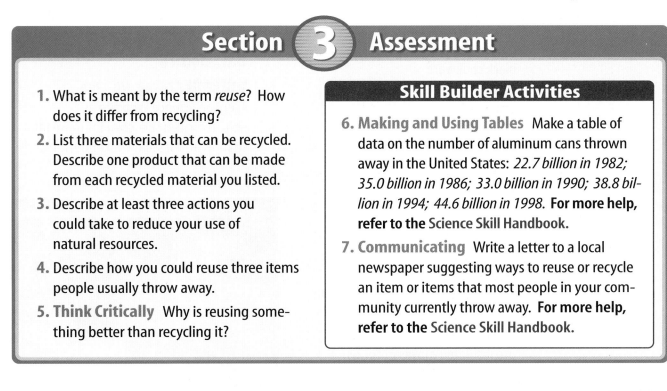

Section ③ Assessment

1. What is meant by the term *reuse?* How does it differ from recycling?

2. List three materials that can be recycled. Describe one product that can be made from each recycled material you listed.

3. Describe at least three actions you could take to reduce your use of natural resources.

4. Describe how you could reuse three items people usually throw away.

5. **Think Critically** Why is reusing something better than recycling it?

Skill Builder Activities

6. **Making and Using Tables** Make a table of data on the number of aluminum cans thrown away in the United States: *22.7 billion in 1982; 35.0 billion in 1986; 33.0 billion in 1990; 38.8 billion in 1994; 44.6 billion in 1998.* **For more help, refer to the** Science Skill Handbook.

7. **Communicating** Write a letter to a local newspaper suggesting ways to reuse or recycle an item or items that most people in your community currently throw away. **For more help, refer to the** Science Skill Handbook.

Activity *Model and Invent*

Solar Cooking

The disappearance of forests in some places on Earth has made firewood extremely difficult and expensive to obtain. People living in these regions often have to travel long distances or sell some of their food to get firewood. This can be a serious problem for people who may not have much food to begin with. Is there a way they could cook food without using firewood?

Recognize the Problem

Can you design a solar cooker that cooks food without burning fuel?

Thinking Critically

How would you build a cooking device that uses the Sun's energy?

Possible Materials

poster board
cardboard boxes
aluminum foil
string
wire coat hangers
clear plastic sheets
*oven bags
black cookware
thermometer
stopwatch
*timer
glue
tape
scissors
*Alternate materials

Safety Precautions

Be careful when cutting your materials. Your solar cooker will get hot. When handling hot liquids or objects, use insulated gloves or tongs.

Goals

- **Research** designs for solar panel cookers or box cookers.
- **Design** a solar cooker that can be used to cook food.
- **Plan** an experiment to measure the effectiveness of your solar cooker.

Data Source

SCIENCE *Online* Visit the Glencoe Science Web site at **science.glencoe.com** for more information about solar cooker designs.

Planning the Model

1. **Design** a solar cooker. In your Science Journal, explain why you chose this design and draw a picture of it.

2. **Write** a summary explaining how you will measure the effectiveness of your solar cooker. What will you measure? How will you collect and organize your data? How will you present your results?

Check the Model Plans

1. **Compare** your solar cooker design to those of other students.

2. Share your experimental plan with students in your class. Discuss the reasoning behind your plan. Be specific about what you intend to test and how you are going to test it.

3. Make sure your teacher approves your plan before you start working on your model.

Making the Model

1. Using all of the information you have gathered, construct a solar cooker that follows your design.

2. **Test** your design to determine how well it works. Try out a classmate's design. How do the two compare?

Analyzing and Applying Results

1. Combine the results for your entire class and decide which type of solar cooker was most effective. How could you design a more effective solar cooker, based on what you learned from this activity?

2. Do you think your results might have been different if you tested your solar cooker on a different day? Explain. Why might a solar cooker be more useful in some regions of the world than in others?

3. Based on what you've read and the results obtained by you and your classmates, do you think that your solar cooker could boil water? Explain.

4. **Compare** the amount of time needed to cook food with a solar cooker and with more traditional cooking methods. Assuming plenty of sunlight is available, would you prefer to use a solar cooker or a traditional oven? Explain.

*C*ommunicating

Your Data

Prepare a demonstration showing how to use a solar cooker. **Present** your demonstration to another class of students or to a group of friends or relatives. **For more help, refer to the** Science Skill Handbook.

Beauty Plagiarized
by Amitabha Mukerjee

I wandered lonely as a cloud –
Except for a motorboat,
Nary a soul in sight.
Beside the lake beneath the trees,
Next to the barbed wire fence,
There was a picnic table
And beer bottle caps from many years.
A boat ramp to the left,
And the chimney from a power station on the other side,
A summer haze hung in the air,
And the lazy drone of traffic far away.

Crimson autumn of mists and mellow fruitfulness
Blue plastic covers the swimming pools
The leaves fall so I can see
Dark glass reflections in the building
That came up
where the pine cones crunched underfoot . . .

And then it is snow
White lining on trees and rooftops . . .
And through my windshield wipers
The snow is piled dark and grey . . .
Next to my driveway where I check my mail
Little footprints on fresh snow —
A visiting rabbit.

I knew a bank where the wild thyme blew
Over-canopied with luscious woodbine
It is now a landfill —
Fermentation of civilization
Flowers on TV
Hyacinth rose tulip chrysanthemum
Acres of colour
Wind up wrapped in decorous plastic,
In this landfill where oxlips grew. . .

Understanding Literature

Cause and Effect Recognizing cause-and-effect relationships can help you make sense out of what you read. One event causes another event. The second event is the effect of the first event. In the poem, the author describes the causes and effects of pollution and waste. For example, the summer haze is the effect of the "chimney from a power station" as well as the "traffic far away." What other effects do pollution and the use of nonrenewable resources have on nature in the poem?

The poet also makes a connection between the four seasons of the year and the pollution and waste products created by civilization. For example, in the spring, a landfill for dumping garbage replaces a field of wildflowers. Describing four seasons instead of one reinforces the poet's message that the beauty of nature has been stolen, or plagiarized.

Science Connection The poet describes renewable and nonrenewable resources. For example, the lake and tree described in the first verse are renewable resources. The beer bottle caps in the same verse are nonrenewable. Can you identify all the nonrenewable resources in the poem? How does the narrator contribute to the use of fossil fuels and nonrenewable resources?

Linking Science and Writing

Write a Poem Copy the poem you just read in pencil. Change the poem by erasing the lines that describe nonrenewable resources. Add lines to the poem that show how recycling, reusing, and other conservation measures could restore the beauty of the four seasons.

Career Connection

Ecologist

Vandana Shiva is the director of the Research Foundation for Science, Technology, and Natural Resource Policy in Dehradun, India. Shiva had an early appreciation for global natural resources as a daughter of a forester growing up in the Himalayan forest. As an adult she has become involved in the efforts of the Chipko Indian women's movement to save trees in the Himalaya. Her involvement in this movement led to the banning of logging in the Himalayan area above 1,000 meters.

SCIENCE *Online* To learn more about careers in ecology, visit the Glencoe Science Web site at **science.glencoe.com.**

Reviewing Main Ideas

Section 1 Resources

1. Natural resources are the parts of the environment that supply materials needed for the survival of living organisms.

2. Renewable resources are being replaced continually by natural processes.

3. Nonrenewable resources cannot be replaced or are replaced very slowly. *What renewable and nonrenewable resources appear in this photograph?*

4. Energy sources include fossil fuels, wind, solar energy, geothermal energy, hydroelectric power, and nuclear power.

Section 2 Pollution

1. Most air pollution is made up of waste products from the burning of fossil fuels.

2. The greenhouse effect is the warming of Earth by a blanket of heat-reflecting gases in the atmosphere.

3. Water can be polluted by acid rain and by the spilling of oil or other wastes into waterways. *What could have caused this algal bloom?*

4. Solid wastes and hazardous wastes dumped on land or disposed of in landfills can pollute the soil. Erosion can cause the loss of fertile topsoil.

Section 3 The Three Rs of Conservation

1. You can reduce your use of natural resources in many ways.

2. Reusing items is an excellent way to practice conservation.

3. In recycling, materials are changed in some way so that they can be used again. *How does recycling benefit the environment?*

4. Materials that can be recycled include paper, metals, glass, plastics, yard waste, and nonmeat kitchen scraps.

FOLDABLES
Reading & Study Skills

After You Read

Use the Foldable you made at the beginning of the chapter to list ways your community conserves resources under the tabs.

Visualizing Main Ideas

Complete the following concept map on air pollution.
Use the terms smog, acid precipitation, *and* ozone depletion.

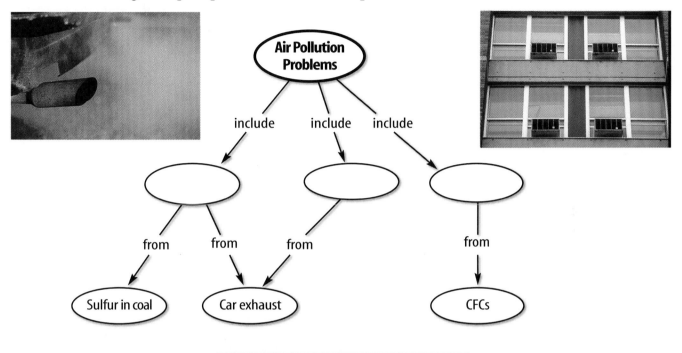

Air Pollution Problems

include   include   include

from   from   from   from

Sulfur in coal   Car exhaust   CFCs

Vocabulary Review

Vocabulary Words

a. acid precipitation
b. erosion
c. fossil fuel
d. geothermal energy
e. greenhouse effect
f. hazardous waste
g. hydroelectric power
h. natural resource
i. nonrenewable resource
j. nuclear energy
k. ozone depletion
l. petroleum
m. pollutant
n. recycling
o. renewable resource

Using Vocabulary

Explain the differences in the vocabulary words given below. Then explain how the words are related. Use complete sentences in your answers.

1. fossil fuel, petroleum
2. erosion, pollutant
3. ozone depletion, acid precipitation
4. greenhouse effect, fossil fuels
5. hazardous wastes, nuclear energy
6. hydroelectric power, fossil fuels
7. acid precipitation, fossil fuels
8. ozone depletion, pollutant
9. recycle, nonrenewable resources
10. geothermal energy, fossil fuels

Checking Concepts

Choose the word or phrase that best answers the question.

1. Which of the following is a fossil fuel?
   A) wood
   B) oil
   C) nuclear power
   D) photovoltaic cell

2. Of the following, which is considered a renewable resource?
   A) coal
   B) oil
   C) sunlight
   D) aluminum

3. Which energy resource uses heat from below Earth's crust?
   A) solar energy
   B) geothermal energy
   C) hydroelectric energy
   D) photovoltaic energy

4. An architect wants to design a solar house in the northern hemisphere. For maximum warmth, which side of the house should have the most windows?
   A) north
   B) south
   C) east
   D) west

5. Which of the following contributes to ozone depletion?
   A) carbon dioxide
   B) radon
   C) CFCs
   D) carbon monoxide

6. What is a substance that contaminates the environment called?
   A) acid rain
   B) pollution
   C) pollutant
   D) ozone

7. If there were no greenhouse effect in Earth's atmosphere, which of the following statements would be true?
   A) Earth would be much hotter.
   B) Earth would be much colder.
   C) The temperature of Earth would be the same.
   D) The polar ice caps would melt.

8. What is the process in which glass bottles are crushed, melted, and shaped into new bottles?
   A) reuse
   B) recycling
   C) reduction
   D) coal

9. Which of the following can change solar energy into electricity?
   A) photovoltaic cells
   B) smog
   C) nuclear power plants
   D) geothermal power plants

10. What is a radioactive gas that can cause indoor air pollution?
    A) ozone
    B) carbon dioxide
    C) radon
    D) chlorofluorocarbons (CFCs)

Thinking Critically

11. Why do burning wood and burning fossil fuels produce similar pollutants?

12. Which would make a better location for a solar power plant—a polar region (left) or a desert region(right)? Why?

13. Why is it beneficial to grow a different crop on soil after the major crop has been harvested?

14. Is garbage a renewable resource? Why or why not?

15. Solar, nuclear, wind, water, and geothermal energy are alternatives to fossil fuels. Are they all renewable? Why or why not?

Developing Skills

16. Drawing Conclusions Would you save more energy by recycling or reusing a plastic bag?

17. Recognizing Cause and Effect Forests use large amounts of carbon dioxide during photosynthesis. How might cutting down a large percentage of Earth's forests affect the greenhouse effect?

18. Making and Using Graphs Make a bar graph of the following data.

Estimated Recycling Rates	
Item	Percent Recycled
Aluminum cans	60
Glass beverage bottles	31
Plastic soft-drink containers	37
Newsprint	56
Magazines	23

19. Forming Hypotheses Form a hypothesis about why Americans throw away more aluminum cans each year.

20. Comparing and Contrasting Compare and contrast contour farming, terracing, strip cropping, and no-till farming.

Performance Assessment

21. Poster Create a poster to illustrate and describe three things students at your school can do to conserve natural resources.

TECHNOLOGY

Go to the Glencoe Science Web site at **science.glencoe.com** or use the **Glencoe Science CD-ROM** for additional chapter assessment.

THE PRINCETON REVIEW   Test Practice

Some of Earth's important natural resources are pictured in the two boxes below.

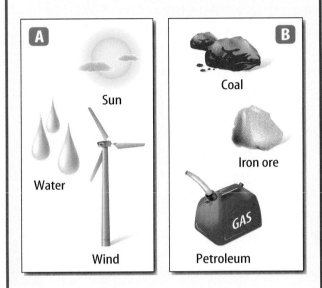

A — Sun, Water, Wind
B — Coal, Iron ore, Petroleum (GAS)

Review the pictures and answer the following questions.

1. Which of the following is a major characteristic of all of the natural resources shown in Box B?
   A) They are made and used by humans on a regular basis.
   B) They are substances made out of metal.
   C) They are materials that occur naturally in a solid form not a liquid form.
   D) They are used up by humans faster than nature can replace them.

2. The resources in Box A are different from the resources in Box B because only the resources in Box A are _____
   F) easily and quickly replaced by nature.
   G) easily and quickly created by humans.
   H) never made by nature.
   J) never used by humans.

5

Conserving Life

Every year since you can remember, your family has driven to this remote area to spend a week camping. The fresh air, the lack of city noise, and the views remind you that your modern lifestyle comes at a cost. Even in the few years that you have been coming here, you notice that there are fewer birds and other animal species than there used to be. In this chapter, you'll learn about biodiversity, endangered and threatened species, and pollution. You'll see that enjoyment of nature is only one of many important reasons for conserving life.

What do you think?

Science Journal Look at the picture below with a classmate. Discuss what is happening. Here's a hint: *This is one method that biologists use to help preserve habitat for wildlife.* Write your answer in your Science Journal.

Do human actions affect the number of species present in an ecosystem? What happens to organisms that are pushed out of an area when the environment changes?

Recognize environmental differences

1. Using string or tape, mark off 1-m × 1-m area of lawn, sports field, or sidewalk.

2. Count and record the different species of plants and animals present in the sample area. Don't forget to include insects or birds that fly over the area, or organisms found by gently probing into the soil.

3. Repeat steps 1 and 2 in a partially wooded area, in a weedy area, or at the edge of a pond or stream.

Observe

Make notes about your observations in your Science Journal. What kinds of human actions could have affected the environment of each sample area?

Before You Read

Making a Know-Want-Learn Study Fold Make the following Foldable to help identify what you already know and what you want to know about biodiversity.

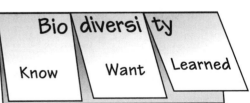

1. Place a sheet of paper in front of you so the long side is at the top. Fold the paper in half from top to bottom.

2. Fold both sides in. Unfold the paper.

3. Through the top thickness of paper, cut along each of the fold lines to the top fold, forming three tabs. Label the tabs *Know, Want,* and *Learned* as shown and then write *Biodiversity* across the front of the paper.

4. Before you read the chapter, write what you know and what you want to know about biodiversity under the tabs.

5. As you read the chapter, add to or correct what you have written under the tabs.

Biodiversity

As You Read

***What* You'll Learn**

- **Define** biodiversity.
- **Explain** why biodiversity is important in an ecosystem.
- **Identify** factors that limit biodiversity in an ecosystem.

Vocabulary

biodiversity
extinct species
endangered species
threatened species
introduced species
native species
acid rain
ozone depletion

***Why* It's Important**

Knowledge of biological diversity can lead to strategies for preventing the loss of species.

Figure 1
The forest has more species and is richer in biodiversity than a wheat field.

The Variety of Life

Imagine walking through a forest ecosystem like the one shown in **Figure 1.** Trees, shrubs, and small plants are everywhere. You see and hear squirrels, birds, and insects. You might notice a snake or mushrooms. Hundreds of species live in this forest. Now, imagine walking through a wheat field. You see only a few species—wheat plants, insects, and weeds. The forest contains more species than the wheat field does. The forest has a higher biological diversity, or biodiversity. **Biodiversity** refers to the variety of life in an ecosystem.

Measuring Biodiversity The common measure of biodiversity is the number of species that live in an area. For example, a coral reef can be home to thousands of species including corals, fish, algae, sponges, crabs, and worms. A coral reef has greater biodiversity than the shallow waters that surround it. Before deep-sea exploration, it was thought that few organisms could live in dark, deep-sea waters. Although it is estimated that the number of organisms living there is less than the number of organisms on a coral reef, it is known that the biodiversity of deep-sea waters is as great as that of a coral reef.

Figure 2
This map shows the number of mammal species found in three North American countries. In general, biodiversity increases as you get closer to the equator.

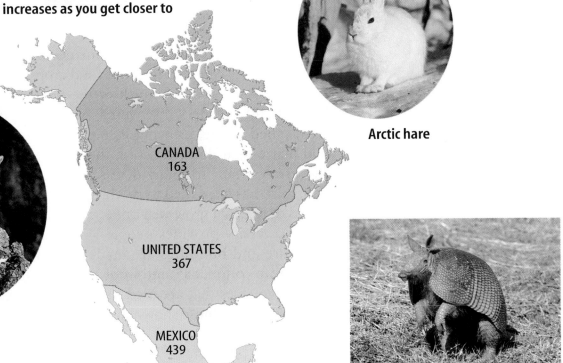

Arctic hare

Mountain lion

Armadillo

Differences in Biodiversity Biodiversity tends to increase as you move toward the equator because temperatures tend to be warmer. For example, Costa Rica is a Central American country about the size of West Virginia. Yet it is home to as many bird species as there are in the United States and Canada combined. **Figure 2** compares mammal biodiversity in three North American countries. Ecosystems with the highest biodiversity usually have warm, moist climates. In fact, tropical regions contain two thirds of all of Earth's land species.

✔ **Reading Check** *What kind of climate usually has a high biodiversity?*

Why is biodiversity important?

Would you rather have a summer picnic on a paved parking lot or on cool grass under a shade tree near a stream? Many people find pleasure in nature's biodiversity. Perhaps you like to visit scenic areas to hike, swim, relax, or enjoy the views. Many painters, writers, and musicians find inspiration for their work by spending time outdoors. However, beauty and pleasure are not the only reasons why biodiversity is important.

Humans Need Biodiversity What foods do you like to eat? If you eat meat or fish, your meals probably include beef, chicken, pork, tuna, shrimp, or clams. Rice, peanuts, green beans, strawberries, or carrots are plant foods you might enjoy. Eating a variety of foods is a good way to stay healthy. Hundreds of species help feed the human population all around the world.

Biodiversity can help improve food crops. During the 1970s, plant diseases wiped out much of the corn crop in the United States. Plant breeders discovered a small population of wild corn, called maize, growing in Mexico. By crossbreeding the wild maize with domestic corn, they developed strains of corn that are resistant to many diseases. Crossbreeding with wild plants also has been used to produce disease-resistant wheat and strains of rice that grow well under drought conditions.

Biodiversity provides people with many useful materials. Furniture and buildings are made from wood and bamboo. Fibers from cotton, flax, and wool are woven into clothing. Most of the medicines that are used today originally came from wild plants, including those shown in **Figure 3.**

✔ Reading Check *Why is biodiversity important to people?*

Figure 3
Although most medicines are made in factories, many originally came from wild plants.

A The velvet bean plant is the original source of L-dopa, a drug used to treat Parkinson's disease.

B An extract from the bark of the Pacific yew was traditionally used to treat arthritis and rheumatism. It is now known that the bark contains taxol, which is a treatment for some types of cancer.

C The bark of the cinchona tree contains quinine, which has been used to cure malaria in millions of people.

D Ephedra contains a decongestant used in the treatment of asthma and other respiratory ailments.

Maintaining Stability Forests usually contain many different kinds of plants. If one type of plant disappears, the forest still exists. Biodiversity allows for stability in an ecosystem.

A loss of biodiversity can weaken an ecosystem. Consider what might happen if a disease infects one grapevine in a vineyard, as shown in **Figure 4.** The vines grow close together, and a disease could move easily from one plant to the next. Soon, the entire vineyard could be infected. Many farmers and gardeners have found that planting alternate rows of different crops can help prevent disease and reduce or eliminate the need for pesticides.

Figure 4
Disease has spread from one grapevine to another in this vineyard.

Math Skills Activity

Analyzing Data for Biodiversity

Example Problem

Temperate rain forest ecosystems in Alaska and California are similar. Which one has the higher mammal and bird biodiversity?

Biodiversity in Temperate Rain Forest Ecosystems

Number of Species	California	Alaska
Mammals	49	38
Birds	94	79
Totals	143	117

Solution

1. *This is what you know:*
   The coastal temperate rain forest of Alaska has 38 mammal species and 79 permanent bird species. The coastal temperate rain forest of California has 49 mammal species and 94 permanent bird species.

2. *This is what you need to find:*
   Which of these ecosystems has the higher bird and mammal biodiversity?

3. *This is the data table you need to create:*
   Calculate the total number of bird and mammal species in each ecosystem.

4. *Compare the totals to determine which ecosystem has the larger number of bird and mammal species:*
   The ecosystem with the larger number of species has a higher bird and mammal biodiversity.

   #### Practice Problem

   Compare the biodiversity of the temperate rain forest ecosystems of Oregon and British Columbia. The Oregon ecosystem has 55 mammal species and 99 permanent bird species. The British Columbia ecosystem has 68 mammal species and 116 permanent bird species.

For more help, refer to the Math Skills Handbook.

What reduces biodiversity?

Flocks of thousands of passenger pigeons used to fly across the skies of North America. Few people today have ever seen one of these birds. The passenger pigeon, shown in **Figure 5,** has been extinct for almost 100 years. An **extinct species** is a species that was once present on Earth but has died out.

Earth Science
INTEGRATION

Extinction is a normal part of nature. The fossil record shows that many species have become extinct since life appeared on Earth. Extinctions can be caused by competition from other species or by changes in the environment. A mass extinction that occurred about 65 million years ago wiped out almost two thirds of all species living on Earth, including the dinosaurs. This extinction, shown on the graph in **Figure 6,** occurred in the Mesozoic Era. It might have been caused by a huge meteorite that crashed into Earth's surface. Perhaps the impact filled the atmosphere with dust and ash that blocked sunlight from reaching Earth's surface. This event might have caused climate changes that many species could not survive. Mass extinctions eventually are followed by the appearance of new species that take advantage of the suddenly empty environment. After the dinosaurs disappeared, many new species of mammals appeared on Earth.

✔ **Reading Check** *What are some causes of extinction?*

Loss of Species Not everyone agrees about the reason for the extinction of the dinosaurs. One thing is clear—human actions had nothing to do with it. Dinosaurs were extinct millions of years before humans were on Earth. Today is different. The rate of extinctions appears to be rising. From 1980 to 2000, close to 40 species of plants and animals in the United States became extinct. It is estimated that hundreds, if not thousands, of tropical species became extinct during the same 20-year period. Human activities probably contributed to most of these extinctions. As the human population grows, many more species could be lost.

Figure 5
The passenger pigeon is an extinct species. These North American birds were over-hunted in the 1800s by frontier settlers who used them as a source of food and feathers.

Figure 6
This graph shows the five mass extinctions in Earth's history. The mass extinctions appear as peaks on the graph.

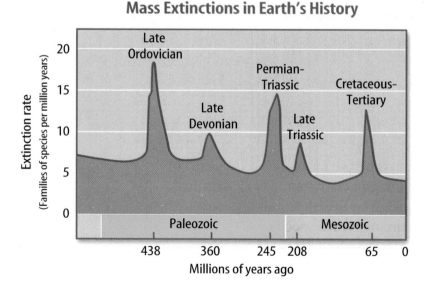

Mass Extinctions in Earth's History

Late Ordovician
Permian-Triassic
Cretaceous-Tertiary
Late Devonian
Late Triassic

Extinction rate
(Families of species per million years)

20
15
10
5
0

Paleozoic
Mesozoic

438   360   245 208   65   0
Millions of years ago

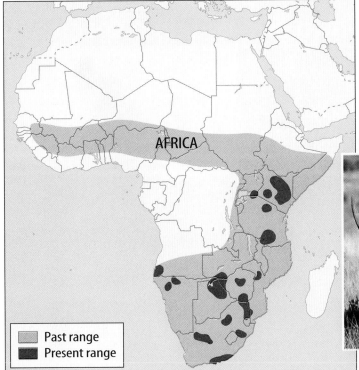

Past range
Present range

Figure 7
The African black rhinoceros is an endangered species, partly because people continue to kill these animals for their horns. As the map shows, this species once lived throughout southern, eastern, and central Africa.

Endangered Species To help prevent extinctions, it is important to identify species that could soon disappear. A species in danger of becoming extinct is classified as an **endangered species.** The African black rhinoceros, shown in **Figure 7,** is endangered. Rhinoceroses are plant eaters. They use their horns to battle each other for territory and to protect themselves against lions and other predators. For centuries, humans have considered rhinoceros horn to be a rare treasure. It is so valuable that poachers continue to hunt and kill these animals, even though selling rhinoceros horn is against international law. In 1970, about 100,000 black rhinoceroses lived in Africa. In the year 2000, fewer than 3,000 lived there.

Threatened Species If a species is likely to become endangered in the near future, it is classified as a **threatened species.** The Australian koala, shown in **Figure 8,** is threatened. People once hunted koalas for their fur. In the 1930s, people realized the koala was in danger. Laws were passed that prohibited the killing of koalas, and the koala populations began to recover. Koalas rely on certain species of Australian eucalyptus trees for food and shelter. Since the 1930 laws were passed, Australia's human population grew and the koala's habitat began to disappear. By the year 2000, nearly two thirds of the koala's habitat had been lost to logging, agriculture, cities, and roads.

Figure 8
Koalas are related to kangaroos and opossums. The loss of their habitat threatens the future of wild koala populations.

Figure 9

The U.S. Fish and Wildlife Service lists more than 1,200 species of animals and plants as being at risk in the United States. Of these, about 20 percent are classified as threatened and 80 percent as endangered.

VERNAL POOL TADPOLE SHRIMP
This endangered species lives in the seasonal freshwater ponds of California's Central Valley. Pollution, urban sprawl, and other forces have destroyed 90 percent of the vernal pools in the valley. Loss of habitat makes the survival of these tiny creatures highly uncertain.

Magnification: 1.5×

CALIFORNIA CONDOR
The endangered condor came close to extinction at the end of the 20th century. Some condors have been raised in captivity and successfully released into the wild.

HALEAKALA SILVERSWORD One of the world's most spectacular plants, this threatened species is making a recovery. The plant shown here is blooming in Hawaii's Haleakala (ha lee ah kuh LAH) crater.

DESERT TORTOISE
The future of the threatened desert tortoise is uncertain. Human development is eroding tortoise habitat in the southwestern United States.

SOUTHERN SEA OTTER The threatened sea otter lives in shallow waters along the Pacific coast of the United States. For centuries, sea otters were hunted for their fur.

Habitat Loss In the Explore Activity at the beginning of this chapter, you probably observed that lawns and sidewalks have a lower biodiversity than sunlit woods or weed-covered lots. When people alter an ecosystem, perhaps by replacing a forest or meadow with pavement or a lawn, the habitats of some species may become smaller or disappear completely. If the habitats of many species are lost, biodiversity might be reduced.

Habitat loss is a major reason why species become threatened or endangered, as shown in **Figure 9,** or extinct. The Lake Erie water snake, shown in **Figure 10B,** is classified as a threatened species because of habitat loss. These snakes live along the rocky shores of islands in Lake Erie. The islands are popular recreation spots. The development of boat docks and buildings in these areas has seriously reduced the amount of habitat available to the snakes. Also, these snakes are often killed by people who mistakenly think they are poisonous. Adult snakes range from 0.5 m to 1.0 m long, and they feed on fish, frogs, and salamanders. Because they have sharp teeth for capturing prey, they can bite. However, they are not dangerous to humans, especially if left undisturbed.

Conservation strategies for protecting the Lake Erie water snake include preserving its habitat by limiting development in some areas. Also, a public education program has been developed to inform people about this predator species and its importance in the Lake Erie ecosystem.

TRY AT HOME
Mini LAB

Demonstrating Habitat Loss

Procedure
1. Put a small piece of **banana** in an **open jar.** Set the jar indoors near a place where food is prepared or fruit is thrown away. Wash your hands when you are finished.
2. Check the jar every few hours. When at least five fruit flies are in it, place a **piece of cloth or stocking** over the top of the jar and secure it with a **rubber band.**
3. Count and record the number of fruit flies in the jar every two days for three weeks.

Analysis
1. Explain why all the flies in the jar eventually die.
2. Use your results to hypothesize why habitat loss can reduce biodiversity.

Figure 10
The Lake Erie water snake has become a threatened species.

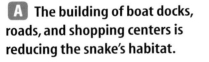
A The building of boat docks, roads, and shopping centers is reducing the snake's habitat.

B These snakes are not dangerous to humans, but many are killed by people who fear snakes.

Divided Habitats Biodiversity can be reduced when a habitat is divided by roads, cities, or farms. Small patches of habitat usually have less biodiversity than large areas. One reason for this is that large animals like mountain lions and grizzly bears require hunting territories that cover hundreds of square kilometers. If their habitat becomes divided, they are forced to move elsewhere.

Small habitat areas also make it difficult for species to recover from a disaster. Suppose a fire destroys part of a forest, and all the salamanders living there are destroyed. Later, after new trees have grown, salamanders from the undamaged part of the forest move in to replace those that were lost. But what if fire destroys a grove of trees surrounded by parking lots and buildings? Trees and salamanders perish. Trees might grow back but the salamanders might never return, because none live in nearby paved areas.

Introduced Species

When species from another part of the world are introduced into an ecosystem, they can have a dramatic effect on biodiversity. An **introduced species** is a species that moves into an ecosystem as a result of human actions. Introduced species often have no competitors or predators in the new area, so their populations grow rapidly. Introduced species can crowd out or consume native species. **Native species** are the original organisms in an ecosystem.

In the early 1800s, European settlers brought goats with them to Santa Catalina Island off the coast of California. The goats overgrazed the land, completely eliminating 48 of the native plant species and exposing patches of bare soil. The bare soil provided a place for hardy, introduced weeds to take root. The weeds crowded out native species. Late in the twentieth century, after goats were removed from the island, some native plant species began to recover. **Figure 11** shows introduced species that have reduced biodiversity in other ecosystems.

Figure 11
Introduced species can reduce or eliminate populations of native species in an ecosystem.

A The ruffe was introduced into the Great Lakes by ships from other parts of the world. It quickly takes over habitat and food sources used by native fish.

B Purple loosestrife was brought to North America from Europe and Asia in the 1800s. Dense patches of loosestrife grow and crowd out the native plants that some animals need for food and shelter.

Pollution

Biodiversity also is affected by pollution of land, water, or air. Soil that is contaminated with oil, chemicals, or other pollutants can harm plants or limit plant growth. Because plants provide valuable habitat for many species, any reduction in plant growth can limit biodiversity.

Water Pollution Water-dwelling organisms are easily harmed by pesticides, chemicals, oil, and other pollutants that contaminate the water. Water pollutants often come from factories, ships, or runoff from roads, lawns, and farms. Waterways also can be polluted when people dispose of wastes improperly. For example, excess water from streets and roads runs into storm drains during rainstorms. This water usually flows untreated into nearby waterways. Storm drains should never be used to dispose of used motor oil, paints, solvents, or other liquid wastes. These pollutants can kill aquatic plants, fish, frogs, insects, and the organisms they depend on for food.

Air Pollution A form of water pollution known as acid rain is caused by air pollution. **Acid rain** forms when sulfur dioxide and nitrogen oxide released by industries and automobiles combine with water vapor in the air. As **Figure 12** shows, acid rain can have serious effects on trees. It washes calcium and other nutrients from the soil, making the soil less fertile. One tree species that is particularly vulnerable to acid rain is the sugar maple. Many sugar maple trees in New England and New York have suffered major damage from acid rain. Acid rain also harms fish and other organisms that live in lakes and streams. Some lakes in Canada have become so acidic that they have lost almost all of their fish species. In the United States, 14 eastern states have acid rain levels high enough to harm aquatic life.

Air pollution from factories, power plants, and automobiles can harm sensitive tissues of many organisms. For example, air pollution can damage the leaves or needles of some trees. This can weaken them and make these trees less able to survive diseases, attacks by insects and other pests, or environmental stresses such as drought or flooding.

Figure 12
Acid rain can damage the leaves and other tissues of trees.

Modeling the Effects of Acid Rain

Procedure

1. Soak **50 mustard seeds** in water and another **50 mustard seeds** in **vinegar** (an acid) for 24 hours.
2. Wrap each group of seeds in a **moist paper towel** and put them into separate, **self-sealing plastic bags**.
3. Seal and label each bag.
4. After 3 days, open each bag and record the number of seeds that show evidence of growth.

Analysis

1. Describe the effect of an acid on mustard seed growth.
2. Explain how acid rain could affect plant biodiversity.

Global Warming Carbon dioxide gas (CO_2) is released into the atmosphere when wood, coal, gas, or any other fuel is burned. People burn large amounts of fuel, and this is contributing to an increase in the percentage of CO_2 in the atmosphere. An increase in CO_2 could raise Earth's average temperature by a few degrees. This average temperature rise, called global warming, might lead to climatic changes that could affect biodiversity. For example, portions of the polar ice caps could melt, causing floods in coastal ecosystems around the world.

Ozone Depletion The atmosphere includes the ozone layer, which protects life on Earth from the Sun's harmful ultraviolet (UV) radiation. The ozone layer consists of ozone gas and is about 15 km to 30 km above Earth's surface. The ozone layer prevents damaging amounts of UV radiation from reaching Earth's surface. Scientists have discovered that the ozone layer is becoming thinner. The thinning of the ozone layer is called **ozone depletion.** This depletion allows increased amounts of UV radiation that can harm living organisms to reach Earth's surface. For humans, it could mean more cases of skin cancer. Ozone depletion occurs over much of Earth. Data taken in the late 1990s indicate that ozone levels over the United States, for example, had decreased by five percent to ten percent since the 1970s.

Health
INTEGRATION

Overexposure to the Sun's radiation can cause health problems such as cataracts, which cloud the lens of the eye, and skin cancer. In your Science Journal, describe ways to limit your exposure to the Sun's damaging radiation.

Section ❶ Assessment

1. What is biodiversity? Give two reasons why it is important.

2. How are endangered species different from extinct species?

3. Why are habitat loss and habitat division threats to biodiversity?

4. In what ways do introduced species threaten biodiversity?

5. **Think Critically** Aluminum, which is commonly found in soil, is toxic to fish. It damages their ability to absorb oxygen through their gills. Acid rain can wash aluminum out of the soil and into nearby waterways. Explain how acid rain could affect biodiversity in a pond.

Skill Builder Activities

6. **Identifying and Manipulating Variables and Controls** Lack of calcium in the soil can damage trees. Design an experiment to test the hypothesis that acid rain is removing calcium from forest soil. **For more help, refer to the Science Skill Handbook.**

7. **Using a Database** Create an endangered species database that includes the following for each species: *common and scientific names, continent(s) on which it is found, habitat description,* and *why it is endangered.* Sort the database to obtain a list of endangered species from each continent. **For more help, refer to the Technology Skill Handbook.**

Activity

Oily Birds

You probably have seen videos of waterbirds covered with oil after an oil spill. Nearly four billion kilograms of oil are spilled from ships or dumped into the ocean each year. Oil floats on water. It can coat the feathers of waterbirds. These birds depend on their feathers for insulation from the cold. Oiled feathers lose their insulating qualities, so oiled birds are exposed to the cold. Oil-coated feathers also prevent the birds from flying and can cause them to drown. Rescue teams try to clean off the oil, but as you'll find out in this activity, this is a difficult task.

What You'll Investigate
How difficult is it to clean oil from bird feathers?

Goal
■ **Experiment** with different methods for cleaning oil from bird feathers.

Materials
beakers (2)	water
bird feathers	cotton balls
(white or light colored)	cotton swabs
vegetable oil or olive oil	sponge
red, blue, and	toothbrush
green food coloring	dish soap
paper towels	

Safety Precautions 🥽 🧤

Procedure

1. Fill your beaker with vegetable oil and add several drops of red, blue, and green food coloring. Submerge your feathers in the oil for several minutes.

2. Lay paper towels on the table. Remove the feathers from the oil after allowing the excess oil to drip back into the beaker. Lay the feathers on the paper towels.

3. Using a paper towel, blot the oil from one feather. Run your fingers over the feather to determine if the oil is gone.

4. Using the cotton swabs and cotton balls, wipe the oil from a second feather. Check the feather for oil when finished.

5. Using a toothbrush, brush the oil from a third feather. Check it for oil when you are finished.

6. Use a sponge to clean the fourth feather. Check it for oil when you are finished.

7. Put several drops of dish soap into a beaker and fill the beaker with water. Soak the fifth feather in the solution. Check the feather for oil after it has soaked for several minutes.

Conclude and Apply

1. **Identify** the method or methods that best removed the oil from the feathers.

2. **Describe** the condition of the feathers after you finished cleaning them. How would the condition of these feathers affect birds?

3. **Infer** why rescuers would be hesitant to use soap when cleaning live birds.

4. **Infer** how oil pollution affects water mammals such as otters and seals.

Communicating
Your Data

Share your results with your classmates. Decide which methods you would use to try to clean a live bird.

Conservation Biology

As You Read

What You'll Learn

- **Identify** several goals of conservation biology.
- **Recommend** strategies to prevent the extinction of species.
- **Explain** how an endangered species can be reintroduced into its original habitat.

Vocabulary

conservation biology
habitat restoration
captive population
reintroduction program

Why It's Important

Conservation biology provides ways to preserve threatened and endangered species.

Protecting Biodiversity

The study of methods for protecting biodiversity is called **conservation biology.** Conservation biologists develop strategies to prevent the continuing loss of members of a species. Conservation strategies must be based on a thorough understanding of the principles of ecology. Because human activities are often the reason why a species is at risk, conservation plans also must take into account the needs of the human population. The needs of humans and of other species often conflict. It can be difficult to develop conservation plans that satisfy both. Conservation biologists must consider law, politics, society, and economics, as well as ecology, when they look for ways to conserve Earth's biodiversity.

The Florida manatee, shown in **Figure 13,** is an endangered species. Manatees are plant-eating mammals that live in shallow water along the coasts of Florida and the Carolinas. These animals swim slowly, occasionally rising to the surface to breathe. They swim below the surface, but not deep enough to avoid the propellers of powerboats. Many manatees have been injured or killed by powerboat propellers. Boaters cannot always tell when manatees are nearby, and it is difficult to enforce speed limits to protect the manatees. Also, the manatee's habitat is being affected by Florida's growing human population. More boat docks are being built, which can add to the number of boats in the water. Water pollution from boats, roads, and coastal cities is also a problem.

✔ **Reading Check** *How do humans affect manatee habitats?*

Figure 13
Conservation biologists are working to protect the endangered Florida manatee by limiting habitat loss, reducing pollution, and encouraging boaters to obey speed limits. These rescued manatees are being rehabilitated at the Columbus Zoo in Ohio.

Figure 14
In the late 1960s, the American bald eagle and American alligator were listed as endangered, so conservation plans were put in place. By the 1990s, their numbers had increased, so they were moved to the threatened species list.

Conservation Biology at Work

Almost every conservation plan has two goals. The first goal is to protect a species from harm. The second goal is to protect the species' habitat. Several conservation strategies can be used to meet these goals.

Legal Protections Laws can be passed to help protect a species and its habitat. In the early 1970s, people became more concerned about the number of species extinctions occurring in the United States. In response to this concern, the U.S. Endangered Species Act of 1973 became a law. This act makes it illegal to harm, collect, harass, or disturb the habitat of any species on the endangered or threatened species lists. The act also prevents the U.S. government from spending money on projects that would harm these species or their habitats. The Endangered Species Act has helped several species, including those shown in **Figure 14,** recover from near extinction.

The United States and other countries worldwide have agreed to work together to protect endangered or threatened species. In 1975, The Convention on International Trade in Endangered Species of Wild Fauna and Flora, known as CITES, came into existence. One of its goals is to protect certain species by controlling or preventing international trade in these species or any part of them, such as elephant ivory. About 5,000 animal species and 25,000 plant species are protected by this agreement.

SCIENCE *Online*

Collect Data For species currently listed as endangered or threatened, visit the Glencoe Science Web site at **science.glencoe.com**. In your Science Journal, write a description of one of these species and explain why it is believed to be in danger.

Yellowstone National Park contains many volcanic features such as geysers, hot springs, and mud volcanoes. The land surrounding these features is fragile. Research what the National Park Service has done to preserve this fragile environment. Communicate to your class what you learn.

Figure 15

A Some wildlife corridors allow animals to pass safely under roads and highways. **B** This map shows the types of habitat where most populations of Florida panthers are found. Notice the major highways that cross these areas.

Habitat Preservation

Even if a species is protected by law, it cannot survive unless its habitat also is protected. Conservation biology often focuses on protecting habitats, or even whole ecosystems. One way to do this is to create nature preserves, such as national parks and protected wildlife areas.

The United States established its first national park—Yellowstone National Park—in 1872. At that time, large animals like grizzly bears, elk, and moose ranged over much of North America. These animals roam across large areas of land in search of food. If their habitat becomes too small, they cannot survive. The grizzly bear, for example, requires large quantities of food each day. To feed itself, a grizzly needs a territory of several hundred square kilometers. Without national parks and wildlife areas, some animals would be far fewer in number than they are today.

Wildlife Corridors

The successful conservation of a species like the grizzly bear requires enormous amounts of land. However, it is not always possible to create large nature preserves. One alternative is to link smaller parks together with wildlife corridors. Wildlife corridors allow animals to move from one preserve to another without having to cross roads, farms, or other areas inhabited by humans. As **Figure 15** shows, wildlife corridors are part of the strategy for saving the endangered Florida panther. A male panther needs a territory of 712 km^2 or more, which is larger than many of the protected panther habitats.

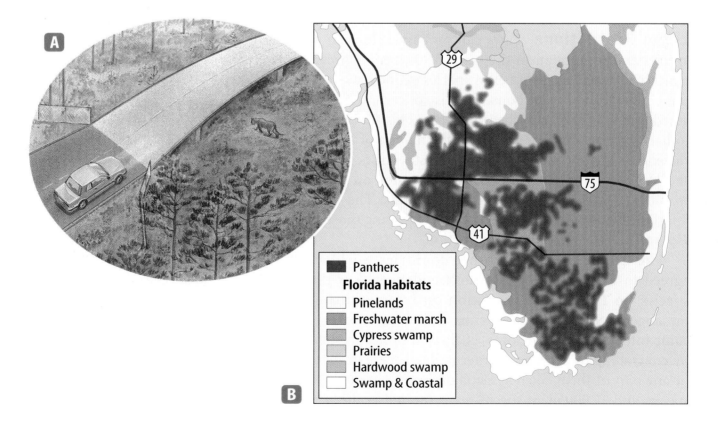

Panthers

Florida Habitats
Pinelands
Freshwater marsh
Cypress swamp
Prairies
Hardwood swamp
Swamp & Coastal

Habitat Restoration Habitats that have been changed or harmed by human activities often can be restored. **Habitat restoration** is the process of taking action to bring a damaged habitat back to a healthy condition.

Rhode Island's Narragansett Bay was once an important fishing area. It supported populations of flounder, bay scallops, blue crabs, and other important food species. Eelgrass, which is an underwater plant that grows in shallow parts of the bay, provides habitat for the young of many of these species. As the human population of Rhode

Island grew, most of the bay's eelgrass beds disappeared. Fish and shellfish populations declined. People who lived in the area knew that something needed to be done. Students in Rhode Island schools have helped restore eelgrass habitat by growing thousands of eelgrass seedlings for transplanting into the bay, as shown in **Figure 16.**

Figure 16
By restoring eelgrass habitat, conservation biologists hope to preserve populations of shellfish and fish in Narragansett Bay.

✔ **Reading Check** *When is habitat restoration used?*

Wildlife Management Preserving or restoring a habitat does not mean that all the species living there are automatically protected from harm. Park rangers, guards, and volunteers often are needed to manage the area. In South Africa, guards patrol wildlife parks to prevent poachers from killing elephants for their tusks. Officials protect the mountain gorillas of Rwanda by limiting the number of visitors allowed to see them. Some wildlife preserves allow no visitors other than biologists who are studying the area.

Hunters and wildlife managers often work together to maintain healthy ecosystems in parks and preserves. People usually are not allowed to hunt or fish in a park unless they purchase a hunting or fishing license. The sale of licenses provides funds for maintaining the wildlife area. It also helps protect populations from overhunting. For example, licenses may limit the number of animals a hunter is allowed to take or may permit hunting only during seasons when a species is not reproducing. Hunting regulations also can help prevent a population from becoming too large for the area.

Captive Populations The Arabian oryx, shown in **Figure 17,** is native to desert lands of the Arabian Peninsula. People hunt the oryx for its horns and hide. Hunting increased during the twentieth century, when four-wheel-drive vehicles became available to hunters. In 1962, to prevent the species from becoming extinct, several oryx were taken to the Phoenix Zoo in Arizona. By 1972, no more wild oryx lived in the desert. A **captive population** is a population of organisms that is cared for by humans. The captive population of oryx in Arizona did well. Their numbers increased, and hundreds of Arabian oryx now live in zoos all over the world.

Figure 17
The Arabian oryx was saved from extinction because a captive population was created before the wild population disappeared.

Keeping endangered or threatened animals in captivity can help preserve biodiversity. It is not ideal, however. It can be expensive to provide proper food, adequate space, and the right kind of care. Also, captive animals sometimes lose their wild behaviors. If that happens, they might not survive if they're returned to their native habitats. The best approach is to preserve the natural habitats of these organisms so they can survive on their own.

SCIENCE Online

Research Visit the Glencoe Science Web site at **science.glencoe.com** to learn more about the Arabian oryx or another reintroduction program. In your Science Journal, write a summary of the program's progress.

Reintroduction Programs In some cases, members of captive populations can be released into the wild to help restore biodiversity. In the 1980s, wildlife managers began reintroducing captive oryx into their desert habitat. **Reintroduction programs** return captive organisms to an area where the species once lived. Programs like this can succeed only if the factors that caused the species to become endangered are removed. In this case, the reintroduced oryx must be protected from illegal hunting.

Plants also can be reintroduced into their original habitats. The island of St. Helena in the South Atlantic was home to the rare St. Helena ebony tree. In the 1500s, European settlers introduced goats to the island. Settlers also cut trees for firewood and timber. Goats overgrazed the land, and native trees were replaced by introduced species. By 1850, the St. Helena ebony was thought to be extinct. In 1980, two small ebony trees were found on the island, high on a cliff. Small branches were cut from these trees and rooted in soil. These cuttings have been used to grow thousands of new ebony trees. The new trees are being replanted all over the island.

Seed Banks Throughout the world, seed banks have been created to store the seeds of many endangered plant species. If any of these species become extinct in the wild, the stored seeds can be used to reintroduce them to their original habitats.

Relocation Reintroduction programs do not always involve captive populations. In fact, reintroductions are most successful when wild organisms are transported to a new area of suitable habitat. The brown pelican, shown in **Figure 18,** was once common along the shores of the Gulf of Mexico. Pelicans eat fish that eat aquatic plants. In the mid-twentieth century, DDT—a pesticide banned in 1972 in the U.S.—was used widely to control insect pests. It eventually ended up in the food that pelicans ate. Because of DDT, the pelican's eggshells were so thin that they would break before the chick inside was ready to hatch. Brown pelicans completely disappeared from Louisiana and most of Texas. In 1971, 50 of these birds were taken from Florida to Louisiana. Since then, the population has grown. In the year 2000, more than 7,000 brown pelicans lived in Louisiana and Texas.

Figure 18
Brown pelicans from Florida were relocated successfully to Louisiana.

 **Reading Check** *What two things led to the recovery of the brown pelican?*

Section ② Assessment

1. What are the differences between ecology and conservation biology?

2. How does the U.S. Endangered Species Act protect species?

3. Why do most conservation plans include habitat preservation strategies?

4. What factors can sometimes make it difficult to reintroduce captive animals into their natural habitats?

5. **Think Critically** As a conservation biologist, you are helping to restore eelgrass beds in Narragansett Bay. What information will you need to prepare an effective conservation plan?

Skill Builder Activities

6. **Recognizing Cause and Effect** Imagine creating a new wildlife preserve in your state. Identify uses of the area that would allow people to enjoy it without harming the ecosystem. What kinds of uses might damage the habitat of species that live there? **For more help, refer to the Science Skill Handbook.**

7. **Communicating** Find information on an endangered or threatened species and a conservation plan that is being used to preserve it. In your Science Journal, write a summary of the conservation plan. **For more help, refer to the Science Skill Handbook.**

Activity

Biodiversity and the Health of a Plant Community

Some plant diseases are carried from plant to plant by specific insects and can spread quickly throughout a garden. If a garden only has plants that are susceptible to one of these diseases, the entire garden can be killed when an infection occurs. One such disease—necrotic leaf spot—can infect impatiens and other garden plants. Could biodiversity help prevent the spread of necrotic leaf spot? A simulation can help you answer this question.

What You'll Investigate

How does biodiversity affect the spread of a plant disease?

Materials

plain white paper	ruler
colored paper (red,	scissors
orange, yellow,	pens
green, blue, black)	die

Safety Precautions

Use care when cutting with scissors.

Goals

- **Record data** on the spread of a plant disease.
- **Compare** the spread of disease in communities with different levels of biodiversity.
- **Predict** a way to increase a crop harvest.

Procedure

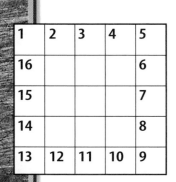

1	2	3	4	5
16				6
15				7
14				8
13	12	11	10	9

1. On a piece of white paper, draw a square that measures 10 cm on each side. This represents a field. In the square, make a grid with five equal rows and five equal columns, as shown. Number the outer cells from 1 through 16, as shown. The inner cells are not numbered.

2. Cut 1.5-cm × 1.5-cm tiles from colored paper. Cut out 25 black tiles, 25 red tiles, 10 orange tiles, 10 yellow tiles, 5 green tiles, and 5 blue tiles. The black tiles represent a plant that has died from a disease. The other colors represent different plant species.

For example, all the red tiles represent one plant species, all the blue tiles are another species, and each remaining color is a different species.

3. There are four rounds in this simulation. For each round, randomly distribute one plant per square as instructed. Roll the die once for each square, proceeding from square 1 to square 2, and so on through square 16.

Round 1—High Biodiversity
Distribute five red, five orange, five yellow, five green, and five blue plants in the field.

Procedures

Round 2—Moderate Biodiversity
Distribute ten orange, ten yellow, and five green plants in the field.

Round 3—Low Biodiversity
Distribute all 25 of the red tiles in the field.

Round 4—Challenge
You want a harvest of as many red plants as possible. Decide how many red plants you will start with and whether you will plant any other species. Strategically place your tiles on the board. Follow steps 3 through 5 of the procedure.

4. To begin, place a black tile over half of square 1. Roll the die and use the Disease Key to see if the plant in square 1 is infected. If it is, cover it with the black tile. If not, remove the black tile. For example, suppose square 1 has an orange plant on it. Place a black tile over half of the square and roll the die. If you roll a 1, 3, 4, or 5, the plant does not get the disease, so remove the black tile. Proceed through all the squares until you have rolled the die 16 times.

5. If two plants of the same species are next to each other, the disease will spread from one to the other. Suppose squares 2 and 3 contain red plants. If you roll a 1 for square 2, the red plants on squares 2 and 3 die. In this case, the die does not need to be rolled for square 3. Similarly, if the inner square next to square 2 also contains a red plant, it dies and should be covered with a black tile.

6. For each round, record in a data table the number of tiles of each species you started with and the number of each species left alive at the end of the round.

Disease Key	
Roll of Die	**Infected Color**
1	Red
2	Orange
3	Yellow
4	Blue
5	Green
6	All colors

Round _____		
Color of Species	**Number at Start**	**Number at End**
Red		
Orange		
Yellow		
Blue		
Green		
TOTAL		

Conclude and Apply

1. Which round had the lowest number of survivors?

2. Why does a high biodiversity help a community survive?

3. Why do organic farmers often grow crops of different species in the same field?

𝒞ommunicating
Your Data

Share your results from Round Four with your classmates. Decide which arrangement and number of red tiles yields the best harvest.

Large areas of Brazil's rain forests are being burned to make way for homes, farms, and ranches.

RAIN FOREST
Troubles

This tropical rain forest in Costa Rica is untouched by human development.

The world's tropical rain forests are disappearing. Here is some background.

Where are Earth's tropical rain forests?

Tropical rain forests are located on both sides of the equator. The average temperature in a rain forest is about 25°C and doesn't vary much between day and night. Rain forests receive, on average, between 200 cm and 400 cm of rain each year.

Why are tropical rain forests in trouble?

Humans destroy as many as 20 million hectares of tropical rain forests each year. Farmers who live in tropical areas cut the trees to sell the wood and farm the land. After a few years, the crops use up the nutrients in the soil and more land must be cleared for farming. The logging and mining industries also contribute to the loss of valuable rain forest resources.

Why are rain forests so special?

Scientists estimate that there are about 30 million plant and animal species on this planet—and at least half of them live in rain forests! Tropical rain forests contain the highest diversity of life on Earth. In one hectare of Brazil's rain forest, you can find 425 different kinds of trees. That's at least ten times the number of trees from comparable areas of Europe and North America. And in one corner of Peru's rain forest, there are 1,300 different types of butterflies.

Even though tropical rain forests are located thousands of miles from the United States, they impact our lives every day. Plants in the forest remove carbon dioxide from the atmosphere and give off oxygen. Some of these plants are already used in medicines, and many more are being studied to determine their usefulness. But if the plants disappear, so might the cures waiting to be discovered.

What can be done to conserve the rain forests?

Conservationists suggest a number of solutions.

- Have governments protect rain forest land by establishing parks and conservation areas.
- Reduce the demand for industrial timber and paper. If people in nontropical areas find ways to cut back on their use of rain forest wood to build homes and make paper, trees will be saved.
- Require industries to use different tree-removal methods that cause less damage. Encourage loggers to plant new trees to replace the ones they cut down.
- Teach farmers alternative farming methods to decrease the damage to rain forest land.
- Offer money or lower taxes to farmers and logging companies for not cutting down trees on rain forest land.

How practical are these solutions?

Carrying out these plans will be difficult. Farmers who clear land to plant crops are concerned about their daily survival. "I have ten children and we must eat," said a farmer from the Ivory Coast, an African nation.

Most scientists believe that at this rate, most of the rain forests will be gone by 2050. The problem extends much farther than the local farmers—the whole world strips rain forest resources in its demand for products such as food ingredients and lumber for building homes. But giving up is not a solution either. "It's going to take a lot to slow down the current rate of destruction," said one conservationist. "We don't have any choice, so we just have to keep churning forward."

Helping to save rain forests is Pedro Sanchez. He specializes in research to find ecologically beneficial ways to farm tropical soils. He says, "Science can be a lot of fun, and I've always liked big challenges. Soil management provides an exciting challenge to find ways to feed the hungry world." And it helps to save the rain forests, too.

The delicate Queen Alexandra's birdwing butterfly lives in the rain forests of Papua New Guinea.

CONNECTIONS **Present** **Prepare a multimedia presentation for an elementary class. Teach the class what the rain forest is, where rain forests are located, and how people benefit from rain forests.**

Online
For more information, visit science.glencoe.com

Section 1 Biodiversity

1. A measure of biodiversity is the number of species present in an ecosystem.

2. In general, biodiversity is greater in warm, moist climates than in cold, dry climates.

3. Extinction occurs when the last member of a species dies. *How was the ancient extinction of the dinosaurs different from the recent extinction of the passenger pigeon?*

4. Habitat loss, pollution, overhunting, and introduced species can cause a species to become threatened or endangered. *In what ways can a road or highway affect biodiversity in an ecosystem?*

Section 2 Conservation Biology

1. Conservation biology is the study of methods for protecting Earth's biodiversity.

2. The Endangered Species Act preserves biodiversity by making it illegal to harm threatened or endangered species.

3. Habitat preservation, habitat restoration, and wildlife management strategies can be used to preserve species. *How does habitat preservation affect the amount of biodiversity?*

4. Reintroduction programs can be used to restore a species to an area where it once lived. *How can captive populations contribute to biodiversity?*

FOLDABLES
Reading & Study Skills

After You Read

Write what you learned about biodiversity under the right tab of your Know-What-Learn Study Fold. Explain the importance of biodiversity.

Visualizing Main Ideas

Complete the following concept map using the following terms:
habitat restoration, captive populations, relocation, and endangered species.

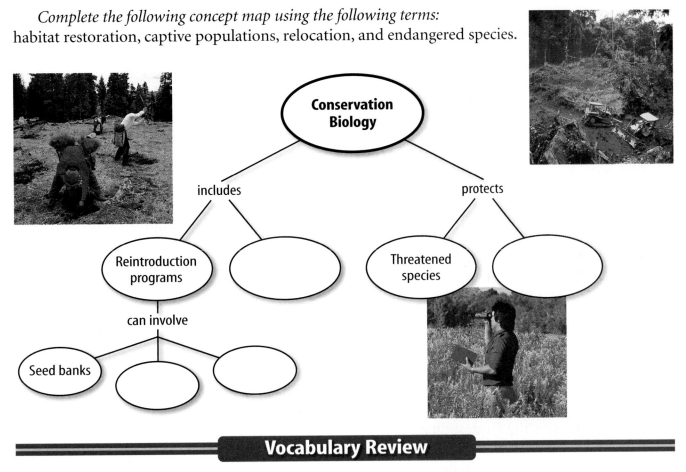

Vocabulary Review

Vocabulary Words

a. acid rain
b. biodiversity
c. captive population
d. conservation biology
e. endangered species
f. extinct species
g. habitat restoration
h. introduced species
i. native species
j. ozone depletion
k. reintroduction program
l. threatened species

Study Tip

Look for science-related news in the newspaper and on television. See how the topics you are studying relate to the real world.

Using Vocabulary

Explain the differences in the vocabulary words given below. Then explain how the words are related.

1. habitat restoration, threatened species

2. endangered species, extinct species

3. biodiversity, conservation biology

4. biodiversity, captive population

5. acid rain, ozone depletion

6. introduced species, endangered species

7. native species, introduced species

8. biodiversity, extinct species

9. reintroduction program, native species

Checking Concepts

Choose the word or phrase that best answers the question.

1. Which state probably has the greatest biodiversity?
   A) Florida
   C) New Hampshire
   B) Maine
   D) West Virginia

2. Jen has a tank with 20 guppies. June has a tank with 5 guppies, 5 mollies, 5 swordtails, and 5 platys. Which of the following is true?
   A) June has more fish than Jen.
   B) June's tank has more biodiversity than Jen's tank.
   C) Jen has more fish than June.
   D) Jen's tank has more biodiversity than June's tank.

3. Which of the following can reduce biodiversity?
   A) divided habitats
   C) introduced species
   B) habitat loss
   D) all of these

4. Which of these is the greatest threat to the Lake Erie water snake?
   A) air pollution
   C) ozone
   B) habitat loss
   D) introduced species

5. What damages the ozone layer?
   A) air pollution
   B) global warming
   C) acid precipitation
   D) biodegradable materials

6. Which of the following gases might contribute to global warming?
   A) oxygen
   C) carbon dioxide
   B) nitrogen
   D) neon

7. What would you call a species that never lived on an island until it was brought there by people?
   A) introduced
   C) native
   B) endangered
   D) threatened

8. Which of the following conservation strategies involves passing laws to protect species?
   A) habitat restoration
   B) legal protections
   C) reintroduction programs
   D) captive populations

9. Which of these organisms is extinct?
   A) passenger pigeon
   C) wheat
   B) African elephant
   D) black rhinoceros

10. The water hyacinth is native to South America. People brought it to the waterways of Florida, where it grows out of control, killing native plants and blocking water flow. Which of the following best describes the water hyacinth?
    A) threatened
    C) introduced
    B) extinct
    D) endangered

Thinking Critically

11. What action(s) could people take to help reduce the types of air pollution that might contribute to global warming?

12. What conservation strategies would most likely help preserve a species whose members are found only in zoos?

13. Four ways to design a road system for a national park are shown below. Which arrangement would best avoid dividing this habitat into pieces?

    a.   b.   c.   d.
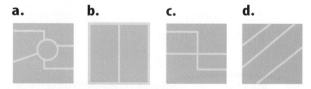

14. Why might the protection of an endangered species create conflict between local residents and conservation biologists?

15. Why is habitat loss the most serious threat to biodiversity?

Developing Skills

16. **Concept Mapping** Draw an events-chain concept map using the following terms: *species, extinct species, endangered species,* and *threatened species.*

17. **Recognizing Cause and Effect** Design a wildlife management plan to allow deer hunting in a state park without damaging the ecosystem.

18. **Making and Using Graphs** Use the data in the table to make a bar graph.

Wildlife Habitat Lost		
Country	**Area Lost (km^2)**	**Area Remaining (km^2)**
Ethiopia	770,700	330,300
Vietnam	265,680	66,420
Indonesia	708,751	746,860

19. **Forming Hypotheses** In divided wildlife areas, large animals have a greater chance of becoming extinct than other organisms do. Suggest a hypothesis to explain why this is true.

20. **Comparing and Contrasting** Explain how legal protections and wildlife management are similar and different.

Performance Assessment

21. **Poster** Use images from magazines to create a display about how humans threaten Earth's biodiversity.

TECHNOLOGY

Go to the Glencoe Science Web site at **science.glencoe.com** or use the **Glencoe Science CD-ROM** for additional chapter assessment.

THE PRINCETON REVIEW Test Practice

A construction team is building a road that runs through an area near two lakes. The team must choose between road plan A and plan B.

Plan A

Plan B

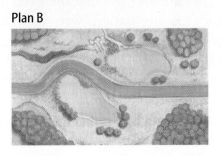

1. A conservation biologist is asked to evaluate how each plan would affect the biodiversity of the ecosystem that includes the two lakes. The biologist urges the road construction team to follow plan A. Why?
   A) It introduces new species.
   B) It does not divide the aquatic habitat into small areas.
   C) It makes the road easier to drive along and view animals.
   D) It helps keep more water in both of the lakes.

2. Which species is least likely to be affected if the road construction team follows plan B?
   F) water snakes   **H)** algae
   G) frogs   **J)** birds

Reading Comprehension

Read the passage. Then read each question that follows the passage. Decide which is the best answer to each question.

Interactions in Ecosystems

Fearing for their safety and the safety of their livestock, early settlers of northern Wisconsin killed the native timber wolves. Timber wolves are a natural predator of white-tailed deer. Over time the deer population increased in size. The available vegetation could not support the deer population. Even though emergency feeding stations were set up, thousands of deer died of starvation. The deer population now is kept down by controlled hunting seasons. In some areas wolves have been reintroduced.

An ecosystem consists of organisms from many different species living together and connected by the flow of energy, nutrients, and matter. Organisms in an ecosystem can be classified as either producers or consumers. Most producers use the Sun's radiant energy and convert it into chemical energy through photo-

synthesis. Consumers take in and use this chemical energy. Herbivores eat producers, carnivores eat other consumers, and omnivores eat producers and consumers. As organisms die, decomposers take in and use the energy in the dead organisms. In doing so, they release nutrients into the soil and carbon dioxide into the air that are used by producers again.

The loss of one species from an ecosystem may lead to the overpopulation or extinction of other species. This loss degrades the ecosystem upon which humans and other organisms depend for clean air, water, and food.

Test-Taking Tip Use the figure to help you visualize the ecosystem that is being described in the passage.

1. Food chains are a way of showing how energy, nutrients, and matter flow through an ecosystem. Which of the following is a food chain of the ecosystem described in the passage?
   A) carnivore, producer, herbivore
   B) producer, herbivore, carnivore
   C) carnivore, producer, decomposer
   D) decomposer, carnivore, herbivore

2. Predators are consumers that capture and eat other consumers. The presence of a predator limits the size of the prey population. This means that food and other resources are less likely to become scarce. What is the predator in this passage?
   F) vegetation
   G) deer
   H) Sun
   J) timberwolf

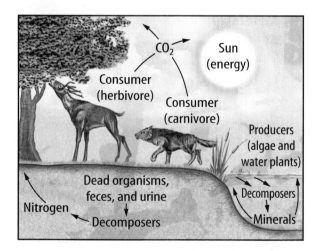

The major components of an ecosystem

Reasoning and Skills

Read each question and choose the best answer.

1. Within an ecosystem there are many populations as well as abiotic factors. Groups of populations that interact within a specific area of an ecosystem are referred to as which of the following?
   A) a habitat
   B) a community
   C) a food web
   D) an atmosphere

Test-Taking Tip   Think about the levels of an ecosystem and how they relate to each other.

World Population

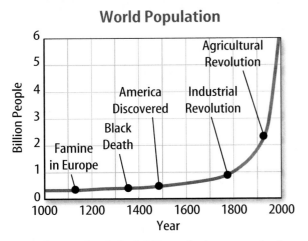

2. Refer to the World Population graph. In which of the following years were the birth rate and the death rate nearly equal?
   F) 1800
   G) 2000
   H) 1200
   J) 1600

Test-Taking Tip   Consider what you know about the effects of birth and death rates on population size.

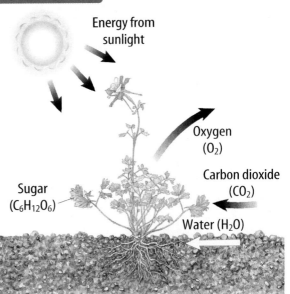

3. The conversion of energy is important to all life on Earth. Some producers use sunlight as an energy source, converting it into chemical energy through photosynthesis. Other producers that live where sunlight does not reach them, can use which of the following as an energy source?
   A) water   **C)** soil
   B) air   **D)** chemicals

Test-Taking Tip   Read about converting energy before answering the question.

Consider this question carefully before writing your answer on a separate sheet of paper.

4. Ecologists are encouraging people to take buses and trains rather than drive their own cars. They say this will help reduce air pollution. How will having more people take public transportation reduce air pollution?

Test-Taking Tip   Imagine how 100 people on a train instead of 100 cars on the highway would affect air pollution.

Student Resources

Student Resources

CONTENTS

Field Guides 156

Biomes Field Guide 156
Waste Management Field Guide 160

Skill Handbooks 164

Science Skill Handbook 164
Organizing Information 164
Researching 164
Evaluating Print and
   Nonprint Sources 164
Interpreting Scientific Illustrations . . 165
Venn Diagram 165
Concept Mapping 165
Writing a Paper 167

Investigating and Experimenting 168
Identifying a Question 168
Forming Hypotheses. 168
Predicting. 168
Testing a Hypothesis 168
Identifying and Manipulating
   Variables and Controls 169
Collecting Data 170
Measuring in SI 171
Making and Using Tables 172
Recording Data 173
Recording Observations 173
Making Models 173
Making and Using Graphs 174

Analyzing and Applying Results 175
Analyzing Results. 175
Forming Operational Definitions . . . 175
Classifying 175
Comparing and Contrasting 176
Recognizing Cause and Effect 176

Interpreting Data. 176
Drawing Conclusions 176
Evaluating Others' Data and
   Conclusions 177
Communicating 177

Technology Skill Handbook 178
Using a Word Processor 178
Using a Database 178
Using an Electronic Spreadsheet 179
Using a Computerized Card Catalog. . . 180
Using Graphics Software. 180
Developing Multimedia
   Presentations 181

Math Skill Handbook 182
Converting Units. 182
Using Fractions 183
Calculating Ratios 184
Using Decimals 184
Using Percentages 185
Using Precision and
   Significant Digits. 185
Solving One-Step Equations 186
Using Proportions. 187
Using Statistics. 188

Reference Handbook 189

A. Safety in the Science Classroom 189
B. Care and Use of a Microscope. 190
C. Diversity of Life. 191

English Glossary 195
Spanish Glossary 199
Index 204

Field GUIDE

Why don't you find polar bears in Florida or palm trees in Alaska? Organisms are limited to certain areas where they can live and survive due to factors such as temperature, amount of rainfall, and type of soil that is found in a region. A biome's boundaries are determined by climate more than anything else. Climate is a way of categorizing temperature extremes and yearly precipitation patterns. Use this field guide to identify some of the world's biomes and to determine which biome you live in.

Key
☐ Temperature range (°C)
▨ Precipitation (cm)

Interpreting Land Biome Climates

The following graphs represent typical climates in seven different biomes. To read each biome graph, use the following information in the key above. Note how each graph displays temperature range, precipitation levels, and the variation between months.

Biomes

Tundra

Winters in the tundra are long and harsh, and summers are short. There is little precipitation. In the tundra, you find mosses, lichens, grasses, and sedges. The tundra supports weasels, arctic foxes, arctic hares, snowy owls, and hawks.

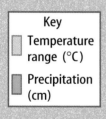

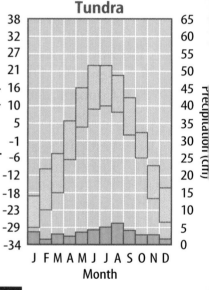

Tundra

Field Activity

Research to find last year's monthly averages for rainfall, high temperature, and low temperature for your area or a nearby city. For further help on last year's averages, visit the Glencoe Science Web site at **science.glencoe.com.** Prepare a graph of data using the example above. Based on your findings, which biome graph most closely matches your data? What biome do you live in? What type of plant and animal life do you expect to find in your biome?

Taiga

Winters in the taiga are cold and severe with much snow. Growing seasons are short. Conifers such as spruces, firs, and larches are common. In the taiga, you find caribou, wolves, moose, bear, ducks, loons, owls and other birds.

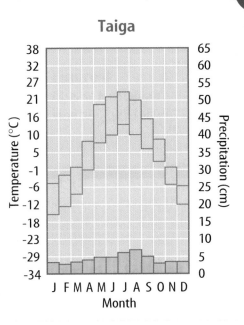

Taiga

Temperate Deciduous Forest

The temperate deciduous forest has cold winters, hot summers, and moderate precipitation. In a temperate deciduous forest, you can see trees such as oak, hickory, and beech, which lose their leaves every autumn. Wolves, deer, bears, small mammals, and many species of birds are common in a temperate deciduous forest.

Temperate Deciduous Forest

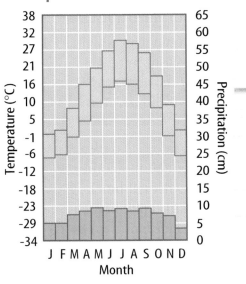

Field GUIDE

Temperate Rain Forest

Temperate Rain forest

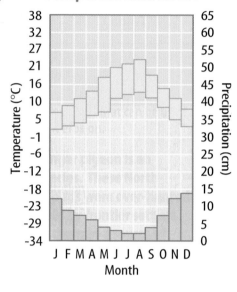

The summers and the winters in the temperate rain forest are mild. Temperatures rarely fall below freezing. The temperate rain forest has heavy precipitation and high humidity. Trees with needlelike leaves, mosses, and ferns are common. Many organisms including salamanders, frogs, black bears, cougars, pileated woodpeckers, and owls make their homes in the temperate rain forest.

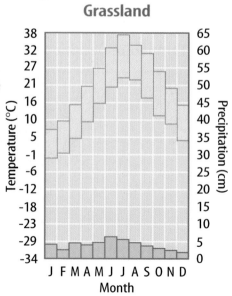

Grassland

There is little precipitation during the grassland's cold winters and hot summers. The plants in the grassland are predominantly grasses although there are also a few trees. The grassland supports grazing animals, wolves, prairies dogs, foxes, ferrets, snakes, lizards, and insects.

Grassland

Desert

Deserts are warm to hot in the daytime and cool in the evening. They receive sparse precipitation throughout the year. Cacti, yuccas, Joshua trees, and bunchgrasses grow in the desert. Small rodents, jackrabbits, birds of prey, and snakes make their homes in the desert.

Desert

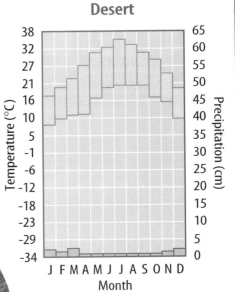

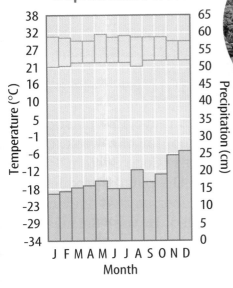

Tropical Rain Forest

Tropical Rain Forest

The tropical rain forest is hot all year with precipitation almost every day. A great diversity of plant species grow in the tropical rain forest. It provides homes for birds, reptiles, insects, monkeys, and sloths.

Field GUIDE

Managing waste properly can reduce the use of resources and prevent pollution. To cut down on waste production and reduce harm to the environment, people can follow the three Rs of waste management: reduce, reuse, and recycle. For example, a 450-g, family-sized box of cereal uses a lot less cardboard than 18 25-g single-serving boxes use. Finding other functions for used items could include donating magazines that you already have read to local hospitals, nursing homes, or even your doctor's office. You can use many products every day that are recyclable. Also, you can help complete the cycle by purchasing items that are made from recycled materials.

Types of Recyclables

Recycled items are either pre- or post-consumer. Pre-consumer items are made of pieces that were left over during manufacturing. Pre-consumer items have never been used by a consumer. Post-consumer items are made of products that have been used, recycled, and made into something new. Cereal boxes, plastic bottles, writing paper, paper towels, and tissues often include post-consumer recycled materials.

Waste Management

Paper

White paper, newspaper, magazines, and telephone books are common paper goods that can be recycled. New products made from these items include newsprint, cardboard, egg cartons, and building materials. Not all glossy and colored papers are recyclable at all recycling centers. As with any recycled material, you should check with your local recycling center for specific instructions about properly sorting your paper goods.

Field Activity

Collect recyclable plastics in your school for one week. Using the Plastics Code System Table, organize the plastic products by code number so that they can be recycled. In your Science Journal, make a bar graph showing the amount and types of plastic you collected.

Plastics

Plastics are difficult products to recycle. Most plastics are composed of complex molecules that tend not to break down easily. Many types of plastics exist, and they often cannot be recycled together. **Table 1** lists the codes that are used to sort plastic items for recycling. Plastic is recycled into lumber, containers, carpet, and other products.

Table 1	The Plastics Code System	
Code	**Material**	
1	PETE	Polyethylene terephthalate
2	HDPE	High-density polyethylene
3	PVC	Polyvinyl chloride
4	LDPE	Low-density polyethylene
5	PP	Polypropylene
6	PS	Polystyrene
7	Other	All others and mixed

Glass

Glass often is separated by color before recycling. Most glass bottles are recyclable, but some glassware such as lightbulbs is too thin to be recycled. New products made from recycled glass include food and beverage containers.

Metals

A variety of metals are recyclable. Steel is the most common material recycled in the United States. Steel is recycled into any item new steel would be used for such as food cans, structural beams for buildings, and parts for automobiles. Aluminum is a commonly recycled material. It is recycled into new beverage cans, lawn chair frames, siding, and cookware. Even precious metals that are used in laboratories or in jewelry such as gold, silver, and platinum are recyclable. Recycling steel saves enough energy to supply one-fifth of the households in the United States (18 million homes) with electricity for one year. A recycled aluminum beverage can uses 95 percent less energy to produce than a can made from new aluminum, which is enough energy to burn a 100W light bulb for three and a half ($3\frac{1}{2}$) hours.

Organic Waste

Organic waste such as yard trimmings, food waste, paper, and wood make up a large percentage of solid waste in the United States. Composting is one way to recycle organic wastes. Food and yard wastes can be placed in compost bins and converted into nutrient rich soil. Mulching is another way to recycle yard waste and wood. In some communities grass clippings, leaves, sticks, and other yard waste can be shredded at local recycling centers to make mulch. Mulch is used to reduce water loss from soil and to keep weeds from growing. As the mulch decays it enriches the soil.

Household Hazardous Wastes

Household Hazardous Wastes (HHWs) include household and car batteries, bleach, household cleaners, paint, paint thinner, motor oil, gasoline, herbicides, pesticides, solvents, and automotive fluids. They contain chemicals that can cause injury or are harmful if used, stored, or discarded improperly. HHWs are marked with a skull and crossbones, caution words or special handling directions. HHWs should not be sent with other garbage to landfills or to incinerators. Water seeps into landfill areas and the toxins could end up in the water supply. Burning may send toxins into the air you breathe. Garbage companies and recycling agencies have free HHW drop-off days or locations.

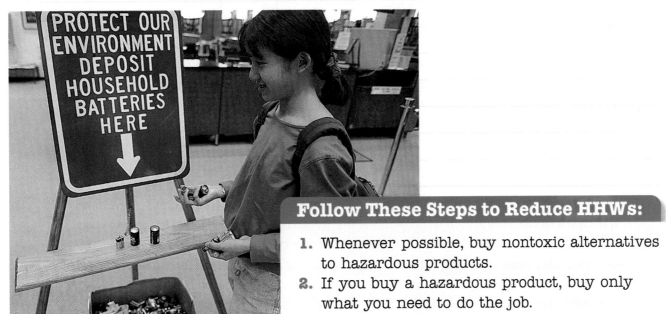

Follow These Steps to Reduce HHWs:

1. Whenever possible, buy nontoxic alternatives to hazardous products.
2. If you buy a hazardous product, buy only what you need to do the job.
3. Before you store leftover products on the shelf, try to find someone who can use them.

Organizing Information

As you study science, you will make many observations and conduct investigations and experiments. You will also research information that is available from many sources. These activities will involve organizing and recording data. The quality of the data you collect and the way you organize it will determine how well others can understand and use it. In **Figure 1,** the student is obtaining and recording information using a microscope.

Putting your observations in writing is an important way of communicating to others the information you have found and the results of your investigations and experiments.

Researching Information

Scientists work to build on and add to human knowledge of the world. Before moving in a new direction, it is important to gather the information that already is known about a subject. You will look for such information in various reference sources. Follow these steps to research information on a scientific subject:

Step 1 Determine exactly what you need to know about the subject. For instance, you might want to find out what happened to local plant life when Mount St. Helens erupted in 1980.

Step 2 Make a list of questions, such as: When did the eruption begin? How long did it last? How large was the area in which plant life was affected?

Step 3 Use multiple sources such as textbooks, encyclopedias, government documents, professional journals, science magazines, and the Internet.

Step 4 List where you found the sources. Make sure the sources you use are reliable and the most current available.

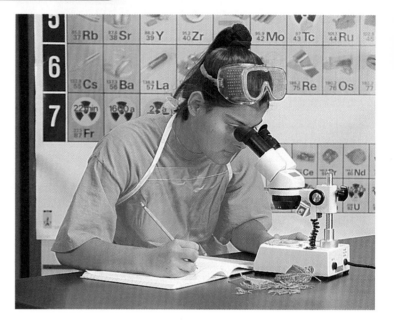

Figure 1
Making an observation is one way to gather information directly.

Evaluating Print and Nonprint Sources

Not all sources of information are reliable. Evaluate the sources you use for information, and use only those you know to be dependable. For example, suppose you want information about the digestion of fats and proteins. You might find two Websites on digestion. One Web site contains "Fat Zapping Tips" written by a company that sells expensive, high-protein supplements to help your body eliminate excess fat. The other is a Web page on "Digestion and Metabolism" written by a well-respected medical school. You would choose the second Web site as the more reliable source of information.

In science, information can change rapidly. Always consult the most current sources. A 1985 source about the human genome would not reflect the most recent research and findings.

Interpreting Scientific Illustrations

As you research a science topic, you will see drawings, diagrams, and photographs. Illustrations help you understand what you read. Some illustrations are included to help you understand an idea that you can't see easily by yourself. For instance, you can't see the bones of a blue whale, but you can look at a diagram of a whale skeleton as labeled in **Figure 2** that helps you understand them. Visualizing a drawing helps many people remember details more easily. Illustrations also provide examples that clarify difficult concepts or give additional information about the topic you are studying.

Most illustrations have a label or a caption. A label or caption identifies the illustration or provides additional information to better explain it. Can you find the caption or labels in **Figure 2?**

Figure 2
A labeled diagram of the skeletal structure of a blue whale

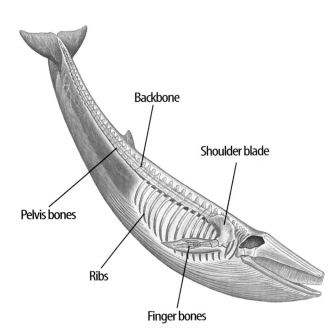

Backbone

Shoulder blade

Pelvis bones

Ribs

Finger bones

Venn Diagram

Although it is not a concept map, a Venn diagram illustrates how two subjects compare and contrast. In other words, you can see the characteristics that the subjects have in common and those that they do not.

The Venn diagram in **Figure 3** shows the relationship between two categories of organisms, plants and animals. Both share some basic characteristics as living organisms. However, there are differences in the ways they carry out various life processes, such as obtaining nourishment, that distinguish one from the other.

Concept Mapping

If you were taking a car trip, you might take some sort of road map. By using a map, you begin to learn where you are in relation to other places on the map.

A concept map is similar to a road map, but a concept map shows relationships among ideas (or concepts) rather than places. It is a diagram that visually shows how concepts are related. Because a concept map shows relationships among ideas, it can make the meanings of ideas and terms clear and help you understand what you are studying.

Overall, concept maps are useful for breaking large concepts down into smaller parts, making learning easier.

Figure 3
A Venn diagram shows how objects or concepts are alike and how they are different.

Plants
(make their own
food, stationary)

Animals
(eat plants or other
animals, move from
place to place)

Alive
(reproduce, grow, and develop)

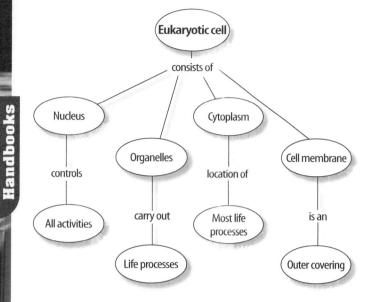

Figure 4
A network tree shows how concepts or objects are related.

Network Tree

Look at the network tree in **Figure 4,** that shows details about a eukaryotic cell. A network tree is a type of concept map. Notice how some words are in ovals while others are written across connecting lines. The words inside the ovals are science terms or concepts. The words written on the connecting lines describe the relationships between the concepts.

When constructing a network tree, write the topic on a note card or piece of paper. Write the major concepts related to that topic on separate note cards or pieces of paper. Then arrange them in order from general to specific. Branch the related concepts from the major concept and describe the relationships on the connecting lines. Continue branching to more specific concepts. Write the relationships between the concepts on the connecting lines until all concepts are mapped. Then examine the network tree for relationships that cross branches, and add them to the network tree.

Events Chain

An events chain is another type of concept map. It models the order of items or their sequence. In science, an events chain can be used to describe a sequence of events, the steps in a procedure, or the stages of a process.

When making an events chain, first find the one event that starts the chain. This event is called the *initiating event.* Then, find the next event in the chain and continue until you reach an outcome. Suppose you are asked to describe the main stages in the growth of a plant from a seed. You might draw an events chain such as the one in **Figure 5.** Notice that connecting words are not necessary in an events chain.

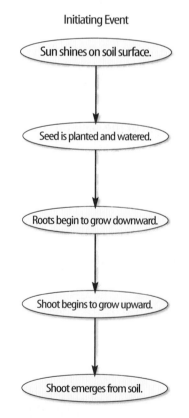

Figure 5
Events chains show the order of steps in a process or event.

Cycle Map A cycle concept map is a specific type of events chain map. In a cycle concept map, the series of events does not produce a final outcome. Instead, the last event in the chain relates back to the beginning event.

You first decide what event will be used as the beginning event. Once that is decided, you list events in order that occur after it. Words are written between events that describe what happens from one event to the next. The last event in a cycle concept map relates back to the beginning event. The number of events in a cycle concept varies but is usually three or more. Look at the cycle map in **Figure 6.**

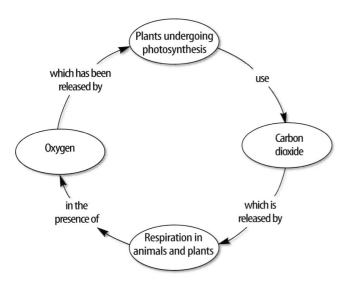

Figure 6
A cycle map shows events that occur in a cycle.

Spider Map A type of concept map that you can use for brainstorming is the spider map. When you have a central idea, you might find you have a jumble of ideas that relate to it but might not clearly relate to each other. The circulatory system spider map in **Figure 7** shows that if you write these ideas outside the main concept, then you can begin to separate and group unrelated terms so they become more useful.

Figure 7
A spider map allows you to list ideas that relate to a central topic but not necessarily to one another.

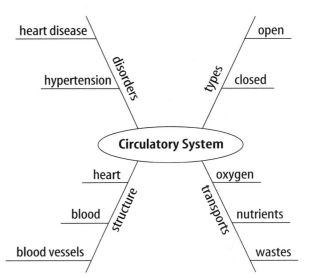

Writing a Paper

You will write papers often when researching science topics or reporting the results of investigations or experiments. Scientists frequently write papers to share their data and conclusions with other scientists and the public. When writing a paper, use these steps.

Step 1 Assemble your data by using graphs, tables, or a concept map. Create an outline.

Step 2 Start with an introduction that contains a clear statement of purpose and what you intend to discuss or prove.

Step 3 Organize the body into paragraphs. Each paragraph should start with a topic sentence, and the remaining sentences in that paragraph should support your point.

Step 4 Position data to help support your points.

Step 5 Summarize the main points and finish with a conclusion statement.

Step 6 Use tables, graphs, charts, and illustrations whenever possible.

Science Skill Handbook

You might say the work of a scientist is to solve problems. When you decide to find out why one corner of your yard is always soggy, you are problem solving, too. You might observe that the corner is lower than the surrounding area and has less vegetation growing in it. You might decide to see if planting some grass will keep the corner drier.

Scientists use orderly approaches to solve problems. The methods scientists use include identifying a question, making observations, forming a hypothesis, testing a hypothesis, analyzing results, and drawing conclusions.

Scientific investigations involve careful observation under controlled conditions. Such observation of an object or a process can suggest new and interesting questions about it. These questions sometimes lead to the formation of a hypothesis. Scientific investigations are designed to test a hypothesis.

Identifying a Question

The first step in a scientific investigation or experiment is to identify a question to be answered or a problem to be solved. You might be interested in knowing why an animal like the one in **Figure 8** looks the way it does.

Figure 8
When you see a bird, you might ask yourself, "How does the shape of this bird's beak help it feed?"

Forming Hypotheses

Hypotheses are based on observations that have been made. A hypothesis is a possible explanation based on previous knowledge and observations.

Perhaps a scientist has observed that bean plants grow larger if they are fertilized than if they are not. Based on these observations, the scientist can make a statement that he or she can test. The statement is a hypothesis. The hypothesis could be: *Fertilizer makes bean plants grow larger.* A hypothesis has to be something you can test by using an investigation. A testable hypothesis is a valid hypothesis.

Predicting

When you apply a hypothesis to a specific situation, you predict something about that situation. First, you must identify which hypothesis fits the situation you are considering. People use predictions to make everyday decisions. Based on previous observations and experiences, you might form a prediction that if fertilizer makes bean plants grow larger, then fertilized plants will yield more beans than plants not fertilized. Someone could use this prediction to plan to grow fewer plants.

Testing a Hypothesis

To test a hypothesis, you need a procedure. A procedure is the plan you follow in your experiment. A procedure tells you what materials to use, as well as how and in what order to use them. When you follow a procedure, data are generated that support or do not support the original hypothesis statement.

For example, suppose you notice that your guppies don't seem as active as usual when your aquarium heater is not working. You wonder how water temperature affects guppy activity level. You decide to test the hypothesis, "If water temperature increases, then guppy activity should increase." Then you write the procedure shown in **Figure 9** for your experiment and generate the data presented in the table below.

Are all investigations alike? Keep in mind as you perform investigations in science that a hypothesis can be tested in many ways. Not every investigation makes use of all the ways that are described on these pages, and not all hypotheses are tested by investigations. Scientists encounter many variations in the methods that are used when they perform experiments. The skills in this handbook are here for you to use and practice.

Identifying and Manipulating Variables and Controls

In any experiment, it is important to keep everything the same except for the item you are testing. The one factor you change is called the independent variable. The factor that changes as a result of the independent variable is called the dependent variable. Always make sure you have only one independent variable. If you allow more than one, you will not know what causes the changes you observe in the dependent variable. Many experiments also have controls—individual instances or experimental subjects for which the independent variable is not changed. You can then compare the test results to the control results.

For example, in the guppy experiment, you made everything the same except the temperature of the water. The glass containers were identical. The volume of aquarium water in each container and beginning water temperature were the same. Each guppy was like the others, as much as possible. In this way, you could be sure that any difference in the number of guppy movements was caused by the temperature change—the independent variable. The activity level of the guppy was measured as the number of guppy movements—the dependent variable. The guppy in the container in which the water temperature was not changed was the control.

Procedure

1. Fill five identical glass containers with equal amounts of aquarium water.
2. Measure and record the temperature of the water in the first container.
3. Heat and cool the other containers so that two have higher and two have lower water temperatures.
4. Place a guppy in each container; count and record the number of movements each guppy makes in 5 minutes.

Figure 9
A procedure tells you what to do step by step.

Number of Guppy Movements		
Container	Temperature (°C)	Movements
1	38	56
2	40	61
3	42	70
4	36	46
5	34	42

Collecting Data

Whether you are carrying out an investigation or a short observational experiment, you will collect data, or information. Scientists collect data accurately as numbers and descriptions and organize it in specific ways.

Observing Scientists observe items and events, then record what they see. When they use only words to describe an observation, it is called qualitative data. For example, a scientist might describe the color of a bird or the shape of a bird's beak as seen through binoculars. Scientists' observations also can describe how much there is of something. These observations use numbers, as well as words, in the description and are called quantitative data. For example, if a particular dog is described as being "furry, yellow, and short-haired," the data are clearly qualitative. Quantitative data for this dog might include "a mass of 14 kg, a height of 46 cm, and an age of 150 days." Quantitative data often are organized into tables. Then, from information in the table, a graph can be drawn. Graphs can reveal relationships that exist in experimental data.

When you make observations in science, you should examine the entire object or situation first, then look carefully for details. If you're looking at a plant, for instance, check general characteristics such as size and overall structure before using a hand lens to examine the leaves and other smaller structures such as flowers or fruits. Remember to record accurately everything you see.

Scientists try to make careful and accurate observations. When possible, they use instruments such as microscopes, metric rulers, graduated cylinders, thermometers, and balances. Measurements provide numerical data that can be repeated and checked.

Sampling When working with large numbers of objects or a large population, scientists usually cannot observe or study every one of them. Instead, they use a sample or a portion of the total number. To *sample* is to take a small, representative portion of the objects or organisms of a population for research. By making careful observations or manipulating variables within a portion of a group, information is discovered and conclusions are drawn that might apply to the whole population.

Estimating Scientific work also involves estimating. To *estimate* is to make a judgment about the size or the number of something without measuring or counting every object or member of a population. Scientists first count the number of objects in a small sample. Looking through a microscope lens, for example, a scientist can count the number of bacterial colonies in the 1-cm^2 frame shown in **Figure 10.** Then the scientist can multiply that number by the number of cm^2 in the petri dish to get an estimate of the total number of bacterial colonies present.

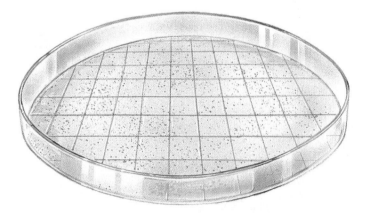

Figure 10
To estimate the total number of bacterial colonies that are present on a petri dish, count the number of bacterial colonies within a 1-cm^2 frame and multiply that number by the number of frames on the dish.

Measuring in SI

The metric system of measurement was developed in 1795. A modern form of the metric system, called the International System, or SI, was adopted in 1960. SI provides standard measurements that all scientists around the world can understand.

The metric system is convenient because unit sizes vary by multiples of 10. When changing from smaller units to larger units, divide by a multiple of 10. When changing from larger units to smaller, multiply by a multiple of 10. To convert millimeters to centimeters, divide the millimeters by 10. To convert 30 mm to centimeters, divide 30 by 10 (30 mm equal 3 cm).

Prefixes are used to name units. Look at the table below for some common metric prefixes and their meanings. Do you see how the prefix *kilo-* attached to the unit *gram* is *kilogram*, or 1,000 g?

Metric Prefixes			
Prefix	Symbol	Meaning	
kilo-	k	1,000	thousand
hecto-	h	100	hundred
deka-	da	10	ten
deci-	d	0.1	tenth
centi-	c	0.01	hundredth
milli-	m	0.001	thousandth

Now look at the metric ruler shown in **Figure 11.** The centimeter lines are the long, numbered lines, and the shorter lines are millimeter lines.

When using a metric ruler, line up the 0-cm mark with the end of the object being measured, and read the number of the unit where the object ends. In this instance it would be 4.5 cm.

Figure 11
This metric ruler shows centimeter and millimeter divisions.

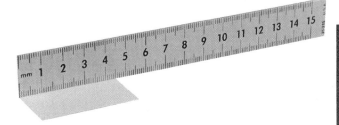

Liquid Volume In some science activities, you will measure liquids. The unit that is used to measure liquids is the liter. A liter has the volume of 1,000 cm^3. The prefix *milli-* means "thousandth (0.001)." A milliliter is one thousandth of 1 L and 1 L has the volume of 1,000 mL. One milliliter of liquid completely fills a cube measuring 1 cm on each side. Therefore, 1 mL equals 1 cm^3.

You will use beakers and graduated cylinders to measure liquid volume. A graduated cylinder, as illustrated in **Figure 12,** is marked from bottom to top in milliliters. This graduated cylinder contains 79 mL of a liquid.

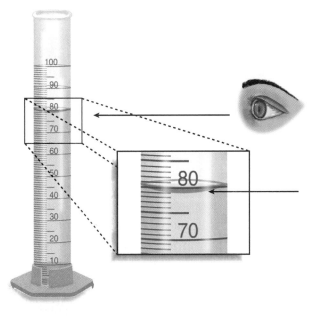

Figure 12
Graduated cylinders measure liquid volume.

Mass Scientists measure mass in grams. You might use a beam balance similar to the one shown in **Figure 13.** The balance has a pan on one side and a set of beams on the other side. Each beam has a rider that slides on the beam.

Before you find the mass of an object, slide all the riders back to the zero point. Check the pointer on the right to make sure it swings an equal distance above and below the zero point. If the swing is unequal, find and turn the adjusting screw until you have an equal swing.

Place an object on the pan. Slide the largest rider along its beam until the pointer drops below zero. Then move it back one notch. Repeat the process on each beam until the pointer swings an equal distance above and below the zero point. Sum the masses on each beam to find the mass of the object. Move all riders back to zero when finished.

Figure 13

A triple beam balance is used to determine the mass of an object.

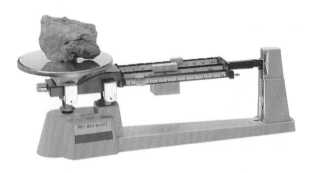

You should never place a hot object on the pan or pour chemicals directly onto the pan. Instead, find the mass of a clean container. Remove the container from the pan, then place the chemicals in the container. Find the mass of the container with the chemicals in it. To find the mass of the chemicals, subtract the mass of the empty container from the mass of the filled container.

Making and Using Tables

Browse through your textbook and you will see tables in the text and in the activities. In a table, data, or information, are arranged so that they are easier to understand. Activity tables help organize the data you collect during an activity so results can be interpreted.

Making Tables To make a table, list the items to be compared in the first column and the characteristics to be compared in the first row. The title should clearly indicate the content of the table, and the column or row heads should tell the reader what information is found in there. The table below lists materials collected for recycling on three weekly pick-up days. The inclusion of kilograms in parentheses also identifies for the reader that the figures are mass units.

Recyclable Materials Collected During Week			
Day of Week	**Paper (kg)**	**Aluminum (kg)**	**Glass (kg)**
Monday	5.0	4.0	12.0
Wednesday	4.0	1.0	10.0
Friday	2.5	2.0	10.0

Using Tables How much paper, in kilograms, is being recycled on Wednesday? Locate the column labeled "Paper (kg)" and the row "Wednesday." The information in the box where the column and row intersect is the answer. Did you answer "4.0"? How much aluminum, in kilograms, is being recycled on Friday? If you answered "2.0," you understand how to read the table. How much glass is collected for recycling each week? Locate the column labeled "Glass (kg)" and add the figures for all three rows. If you answered "32.0," then you know how to locate and use the data provided in the table.

Recording Data

To be useful, the data you collect must be recorded carefully. Accuracy is key. A well-thought-out experiment includes a way to record procedures, observations, and results accurately. Data tables are one way to organize and record results. Set up the tables you will need ahead of time so you can record the data right away.

Record information properly and neatly. Never put unidentified data on scraps of paper. Instead, data should be written in a notebook like the one in **Figure 14.** Write in pencil so information isn't lost if your data gets wet. At each point in the experiment, record your data and label it. That way, your information will be accurate and you will not have to determine what the figures mean when you look at your notes later.

Figure 14
Record data neatly and clearly so it is easy to understand.

Recording Observations

It is important to record observations accurately and completely. That is why you always should record observations in your notes immediately as you make them. It is easy to miss details or make mistakes when recording results from memory. Do not include your personal thoughts when you record your data. Record only what you observe to eliminate bias. For example, when you record that a plant grew 12 cm in one day, you would note that this was the largest daily growth for the week. However, you would not refer to the data as "the best growth spurt of the week."

Making Models

You can organize the observations and other data you collect and record in many ways. Making models is one way to help you better understand the parts of a structure you have been observing or the way a process for which you have been taking various measurements works.

Models often show things that are very large or small or otherwise would be difficult to see and understand. You can study blood vessels and know that they are hollow tubes. The size and proportional differences among arteries, veins, and capillaries can be explained in words. However, you can better visualize the relative sizes and proportions of blood vessels by making models of them. Gluing different kinds of pasta to thick paper so the openings can be seen can help you see how the differences in size, wall thickness, and shape among types of blood vessels affect their functions.

Other models can be devised on a computer. Some models, such as disease control models used by doctors to predict the spread of the flu, are mathematical and are represented by equations.

Making and Using Graphs

After scientists organize data in tables, they might display the data in a graph that shows the relationship of one variable to another. A graph makes interpretation and analysis of data easier. Three types of graphs are the line graph, the bar graph, and the circle graph.

Line Graphs A line graph like in **Figure 15** is used to show the relationship between two variables. The variables being compared go on two axes of the graph. For data from an experiment, the independent variable always goes on the horizontal axis, called the *x*-axis. The dependent variable always goes on the vertical axis, called the *y*-axis. After drawing your axes, label each with a scale. Next, plot the data points.

A data point is the intersection of the recorded value of the dependent variable for each tested value of the independent variable. After all the points are plotted, connect them.

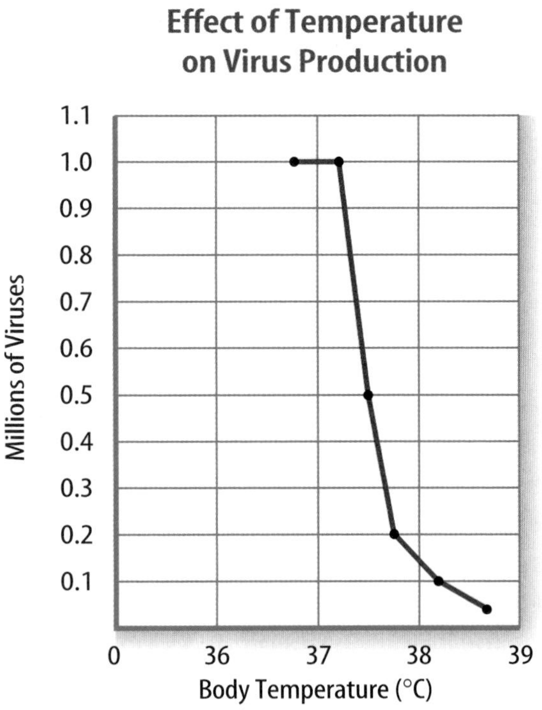

Bar Graphs Bar graphs compare data that do not change continuously. Vertical bars show the relationships among data.

To make a bar graph, set up the *y*-axis as you did for the line graph. Draw vertical bars of equal size from the *x*-axis up to the point on the *y*-axis that represents the value of *x*.

Figure 16

The number of wing vibrations per second for different insects can be shown as a bar graph or circle graph.

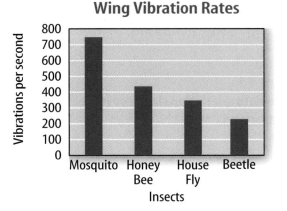

Wing Vibration Rates

Effect of Temperature on Virus Production

Figure 15
This line graph shows the relationship between body temperature and the millions of infecting viruses present in a human body.

Circle Graphs A circle graph uses a circle divided into sections to display data as parts (fractions or percentages) of a whole. The size of each section corresponds to the fraction or percentage of the data that the section represents. So, the entire circle represents 100 percent, one-half represents 50 percent, one-fifth represents 20 percent, and so on.

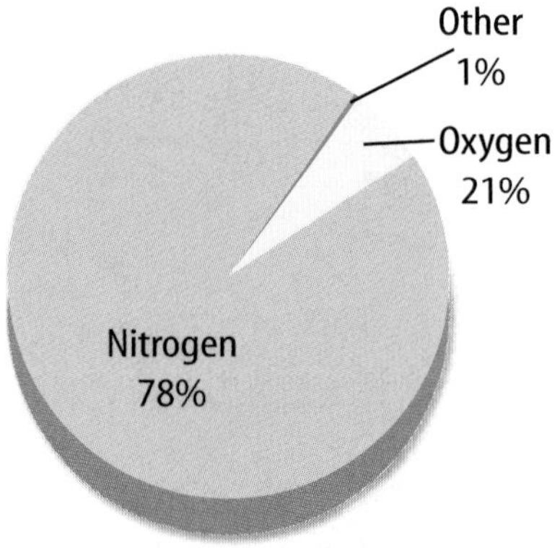

Analyzing Results

To determine the meaning of your observations and investigation results, you will need to look for patterns in the data. You can organize your information in several of the ways that are discussed in this handbook. Then you must think critically to determine what the data mean. Scientists use several approaches when they analyze the data they have collected and recorded. Each approach is useful for identifying specific patterns in the data.

Forming Operational Definitions

An operational definition defines an object by showing how it functions, works, or behaves. Such definitions are written in terms of how an object works or how it can be used; that is, they describe its job or purpose.

For example, a ruler can be defined as a tool that measures the length of an object (how it can be used). A ruler also can be defined as something that contains a series of marks that can be used as a standard when measuring (how it works).

Classifying

Classifying is the process of sorting objects or events into groups based on common features. When classifying, first observe the objects or events to be classified. Then select one feature that is shared by some members in the group but not by all. Place those members that share that feature into a subgroup. You can classify members into smaller and smaller subgroups based on characteristics.

How might you classify a group of animals? You might first classify them by putting all of the dogs, cats, lizards, snakes, and birds into separate groups. Within each group, you could then look for another common feature by which to further classify members of the group, such as size or color.

Remember that when you classify, you are grouping objects or events for a purpose. For example, classifying animals can be the first step in identifying them. You might know that a cardinal is a red bird. To find it in a large group of animals, you might start with the classification scheme mentioned here. You'll locate a cardinal within the red grouping of the birds that you separate from the rest of the animals. A male ruby-throated hummingbird could be located within the birds by its tiny size and the bright red color of its throat. Keep your purpose in mind as you select the features to form groups and subgroups.

Figure 17
Color is one of many characteristics that are used to classify animals.

Science Skill Handbook

Comparing and Contrasting

Observations can be analyzed by noting the similarities and differences between two or more objects or events that you observe. When you look at objects or events to see how they are similar, you are comparing them. Contrasting is looking for differences in objects or events. The table below compares and contrasts the nutritional value of two cereals.

Nutritional Values		
	Cereal A	Cereal B
Calories	220	160
Fat	10 g	10 g
Protein	2.5 g	2.6 g
Carbohydrate	30 g	15 g

Recognizing Cause and Effect

Have you ever gotten a cold and then suggested that you probably caught it from a classmate who had one recently? If so, you have observed an effect and inferred a cause. The event is the effect, and the reason for the event is the cause.

When scientists are unsure of the cause of a certain event, they design controlled experiments to determine what caused it.

Interpreting Data

The word *interpret* means "to explain the meaning of something." Look at the problem originally being explored in an experiment and figure out what the data show. Identify the control group and the test group so you can see whether or not changes in the independent variable have had an effect. Look for differences in the dependent variable between the control and test groups.

These differences you observe can be qualitative or quantitative. You would be able to describe a qualitative difference using only words, whereas you would measure a quantitative difference and describe it using numbers. If there are qualitative or quantitative differences, the independent variable that is being tested could have had an effect. If no qualitative or quantitative differences are found between the control and test groups, the variable that is being tested apparently had no effect.

For example, suppose that three pepper plants are placed in a garden and two of the plants are fertilized, but the third is left to grow without fertilizer. Suppose you are then asked to describe any differences in the plants after two weeks. A qualitative difference might be the appearance of brighter green leaves on fertilized plants but not on the unfertilized plant. A quantitative difference might be a difference in the height of the plants or the number of flowers on them.

Inferring Scientists often make inferences based on their observations. An inference is an attempt to explain, or interpret, observations or to indicate what caused what you observed. An inference is a type of conclusion.

When making an inference, be certain to use accurate data and accurately described observations. Analyze all of the data that you've collected. Then, based on everything you know, explain or interpret what you've observed.

Drawing Conclusions

When scientists have analyzed the data they collected, they proceed to draw conclusions about what the data mean. These conclusions are sometimes stated using words similar to those found in the hypothesis formed earlier in the process.

Conclusions To analyze your data, you must review all of the observations and measurements that you made and recorded. Recheck all data for accuracy. After your data are rechecked and organized, you are almost ready to draw a conclusion such as "Plants need sunlight in order to grow."

Before you can draw a conclusion, however, you must determine whether the data allow you to come to a conclusion that supports a hypothesis. Sometimes that will be the case; other times it will not.

If your data do not support a hypothesis, it does not mean that the hypothesis is wrong. It means only that the results of the investigation did not support the hypothesis. Maybe the experiment needs to be redesigned, but very likely, some of the initial observations on which the hypothesis was based were incomplete or biased. Perhaps more observation or research is needed to refine the hypothesis.

Avoiding Bias Sometimes drawing a conclusion involves making judgments. When you make a judgment, you form an opinion about what your data mean. It is important to be honest and to avoid reaching a conclusion if no supporting evidence for it exists or if it is based on a small sample. It also is important not to allow any expectations of results to bias your judgments. If possible, it is a good idea to collect additional data. Scientists do this all the time.

For example, animal behaviorist Katharine Payne made an important observation about elephant communication. While visiting a zoo, Payne felt the air vibrating around her. At the same time, she also noticed that the skin on an elephant's forehead was fluttering. She suspected that the elephants were generating the vibrations and that they might be using the low-frequency sounds to communicate.

Payne conducted an experiment to record these sounds and simultaneously observe the behavior of the elephants in the zoo. She later conducted a similar experiment in Namibia in southwest Africa, where elephant herds roam. The additional data she collected supported the judgment Payne had made, which was that these low-frequency sounds were a form of communication between elephants.

Evaluating Others' Data and Conclusions

Sometimes scientists have to use data that they did not collect themselves, or they have to rely on observations and conclusions drawn by other researchers. In cases such as these, the data must be evaluated carefully.

How were the data obtained? How was the investigation done? Has it been duplicated by other researchers? Did they come up with the same results? Look at the conclusion, as well. Would you reach the same conclusion from these results? Only when you have confidence in the data of others can you believe it is true and feel comfortable using it.

Communicating

The communication of ideas is an important part of the work of scientists. A discovery that is not reported will not advance the scientific community's understanding or knowledge. Communication among scientists also is important as a way of improving their investigations.

Scientists communicate in many ways, from writing articles in journals and magazines that explain their investigations and experiments, to announcing important discoveries on television and radio, to sharing ideas with colleagues on the Internet or presenting them as lectures.

Skill Handbooks

People who study science rely on computers to record and store data and to analyze results from investigations. Whether you work in a laboratory or just need to write a lab report with tables, good computer skills are a necessity.

Using a Word Processor

Suppose your teacher has assigned a written report. After you've completed your research and decided how you want to write the information, you need to put all that information on paper. The easiest way to do this is with a word processing application on a computer.

A computer application that allows you to type your information, change it as many times as you need to, and then print it out so that it looks neat and clean is called a word processing application. You also can use this type of application to create tables and columns, add bullets or cartoon art to your page, include page numbers, and even check your spelling.

Helpful Hints

- If you aren't sure how to do something using your word processing program, look in the help menu. You will find a list of topics there to click on for help. After you locate the help topic you need, just follow the step-by-step instructions you see on your screen.
- Just because you've spell checked your report doesn't mean that the spelling is perfect. The spell check feature can't catch misspelled words that look like other words. If you've accidentally typed *wind* instead of *wing*, the spell checker won't know the difference. Always reread your report to make sure you didn't miss any mistakes.

Figure 18
You can use computer programs to make graphs and tables.

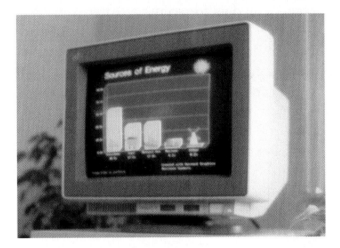

Using a Database

Imagine you're in the middle of a research project busily gathering facts and information. You soon realize that it's becoming more difficult to organize and keep track of all the information. The tool to use to solve information overload is a database. Just as a file cabinet organizes paper records, a database organizes computer records. However, a database is more powerful than a simple file cabinet because at the click of a mouse, the contents can be reshuffled and reorganized. At computer-quick speeds, databases can sort information by any characteristics and filter data into multiple categories.

Helpful Hints

- Before setting up a database, take some time to learn the features of your database software by practicing with established database software.
- Periodically save your database as you enter data. That way, if something happens such as your computer malfunctions or the power goes off, you won't lose all of your work.

Doing a Database Search

When searching for information in a database, use the following search strategies to get the best results. These are the same search methods used for searching Internet databases.

- Place the word *and* between two words in your search if you want the database to look for any entries that have both words. For example, "fox *and* mink" would give you information that mentions both fox and mink.

- Place the word *or* between two words if you want the database to show entries that have at least one of the words. For example "fox *or* mink" would show you information that mentions either fox or mink.

- Place the word *not* between two words if you want the database to look for entries that have the first word but do not have the second word. For example, "canine *not* fox" would show you information that mentions the term *canine* but does not mention the fox.

In summary, databases can be used to store large amounts of information about a particular subject. Databases allow biologists, Earth scientists, and physical scientists to search for information quickly and accurately.

Using an Electronic Spreadsheet

Your science fair experiment has produced lots of numbers. How do you keep track of all the data, and how can you easily work out all the calculations needed? You can use a computer program called a spreadsheet to record data that involve numbers. A spreadsheet is an electronic mathematical worksheet.

Type in your data in rows and columns, just as in a data table on a sheet of paper. A spreadsheet uses simple math to do data calculations. For example, you could add, subtract, divide, or multiply any of the values in the spreadsheet by another number. You also could set up a series of math steps you want to apply to the data. If you want to add 12 to all the numbers and then multiply all the numbers by 10, the computer does all the calculations for you in the spreadsheet. Below is an example of a spreadsheet that records data from an experiment with mice in a maze.

Helpful Hints

- Before you set up the spreadsheet, identify how you want to organize the data. Include any formulas you will need to use.
- Make sure you have entered the correct data into the correct rows and columns.
- You also can display your results in a graph. Pick the style of graph that best represents the data with which you are working.

Figure 19
A spreadsheet allows you to display large amounts of data and do calculations automatically.

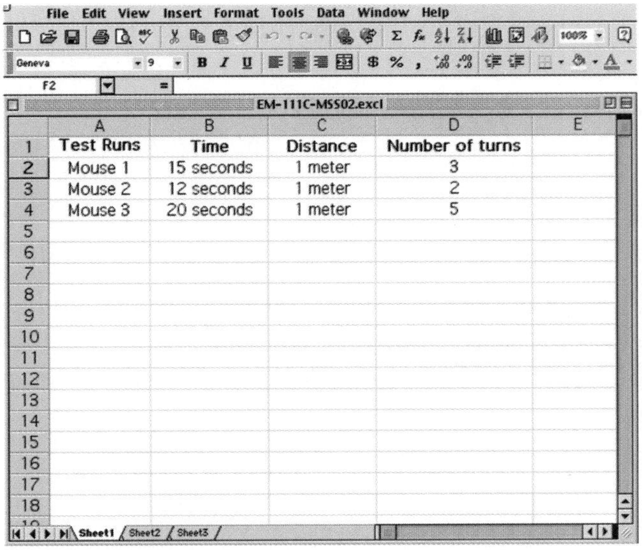

	Test Runs	Time	Distance	Number of turns
1	Test Runs	Time	Distance	Number of turns
2	Mouse 1	15 seconds	1 meter	3
3	Mouse 2	12 seconds	1 meter	2
4	Mouse 3	20 seconds	1 meter	5

Using a Computerized Card Catalog

When you have a report or paper to research, you probably go to the library. To find the information you need in the library, you might have to use a computerized card catalog. This type of card catalog allows you to search for information by subject, by title, or by author. The computer then will display all the holdings the library has on the subject, title, or author requested.

A library's holdings can include books, magazines, databases, videos, and audio materials. When you have chosen something from this list, the computer will show whether an item is available and where in the library to find it.

Helpful Hints

■ Remember that you can use the computer to search by subject, author, or title. If you know a book's author but not the title, you can search for all the books the library has by that author.

■ When searching by subject, it's often most helpful to narrow your search by using specific search terms, such as *and, or,* and *not.* If you don't find enough sources, you can broaden your search.

■ Pay attention to the type of materials found in your search. If you need a book, you can eliminate any videos or other resources that come up in your search.

■ Knowing how your library is arranged can save you a lot of time. The librarian will show you where certain types of materials are kept and how to find specific holdings.

Using Graphics Software

Are you having trouble finding that exact piece of art you're looking for? Do you have a picture in your mind of what you want but can't seem to find the right graphic to represent your ideas? To solve these problems, you can use graphics software. Graphics software allows you to create and change images and diagrams in almost unlimited ways. Typical uses for graphics software include arranging clip art, changing scanned images, and constructing pictures from scratch. Most graphics software applications work in similar ways. They use the same basic tools and functions. Once you master one graphics application, you can use any other graphics application relatively easily.

Figure 20
Graphics software can use your data to draw bar graphs.

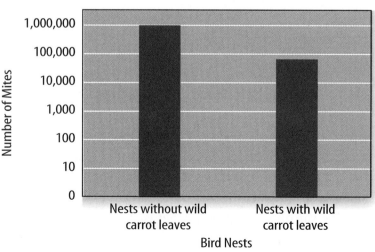

Number of Mites per Bird Nest

Figure 21
Graphics software can use your data to draw circle graphs.

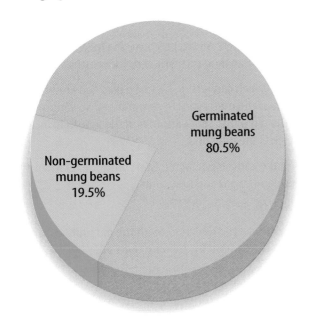

Germinated
mung beans
80.5%

Non-germinated
mung beans
19.5%

Helpful Hints
- As with any method of drawing, the more you practice using the graphics software, the better your results will be.
- Start by using the software to manipulate existing drawings. Once you master this, making your own illustrations will be easier.
- Clip art is available on CD-ROMs and the Internet. With these resources, finding a piece of clip art to suit your purposes is simple.
- As you work on a drawing, save it often.

Developing Multimedia Presentations

It's your turn—you have to present your science report to the entire class. How do you do it? You can use many different sources of information to get the class excited about your presentation. Posters, videos, photographs, sound, computers, and the Internet can help show your ideas.

First, determine what important points you want to make in your presentation. Then, write an outline of what materials and types of media would best illustrate those points. Maybe you could start with an outline on an overhead projector, then show a video, followed by something from the Internet or a slide show accompanied by music or recorded voices. You might choose to use a presentation builder computer application that can combine all these elements into one presentation. Make sure the presentation is well constructed to make the most impact on the audience.

Figure 22
Multimedia presentations use many types of print and electronic materials.

Helpful Hints
- Carefully consider what media will best communicate the point you are trying to make.
- Make sure you know how to use any equipment you will be using in your presentation.
- Practice the presentation several times.
- If possible, set up all of the equipment ahead of time. Make sure everything is working correctly.

Math Skill Handbook

Use this Math Skill Handbook to help solve problems you are given in this text. You might find it useful to review topics in this Math Skill Handbook first.

Converting Units

In science, quantities such as length, mass, and time sometimes are measured using different units. Suppose you want to know how many miles are in 12.7 km.

Conversion factors are used to change from one unit of measure to another. A conversion factor is a ratio that is equal to one. For example, there are 1,000 mL in 1 L, so 1,000 mL equals 1 L, or:

$$1{,}000 \text{ mL} = 1 \text{ L}$$

If both sides are divided by 1 L, this equation becomes:

$$\frac{1{,}000 \text{ mL}}{1 \text{ L}} = 1$$

The **ratio** on the left side of this equation is equal to 1 and is a conversion factor. You can make another conversion factor by dividing both sides of the top equation by 1,000 mL:

$$1 = \frac{1 \text{ L}}{1{,}000 \text{ mL}}$$

To **convert units,** you multiply by the appropriate conversion factor. For example, how many milliliters are in 1.255 L? To convert 1.255 L to milliliters, multiply 1.255 L by a conversion factor.

Use the **conversion factor** with new units (mL) in the numerator and the old units (L) in the denominator.

$$1.255 \text{ L} \times \frac{1{,}000 \text{ mL}}{1 \text{ L}} = 1{,}255 \text{ mL}$$

The unit L divides in this equation, just as if it were a number.

Example 1 There are 2.54 cm in 1 inch. If a meterstick has a length of 100 cm, how long is the meterstick in inches?

Step 1 Decide which conversion factor to use. You know the length of the meterstick in centimeters, so centimeters are the old units. You want to find the length in inches, so inch is the new unit.

Step 2 Form the conversion factor. Start with the relationship between the old and new units.

$$2.54 \text{ cm} = 1 \text{ inch}$$

Step 3 Form the conversion factor with the old unit (centimeter) on the bottom by dividing both sides by 2.54 cm.

$$1 = \frac{2.54 \text{ cm}}{2.54 \text{ cm}} = \frac{1 \text{ inch}}{2.54 \text{ cm}}$$

Step 4 Multiply the old measurement by the conversion factor.

$$100 \text{ cm} \times \frac{1 \text{ inch}}{2.54 \text{ cm}} = 39.37 \text{ inches}$$

The meterstick is 39.37 inches long.

Example 2 There are 365 days in one year. If a person is 14 years old, what is his or her age in days? (Ignore leap years)

Step 1 Decide which conversion factor to use. You want to convert years to days.

Step 2 Form the conversion factor. Start with the relation between the old and new units.

$$1 \text{ year} = 365 \text{ days}$$

Step 3 Form the conversion factor with the old unit (year) on the bottom by dividing both sides by 1 year.

$$1 = \frac{1 \text{ year}}{1 \text{ year}} = \frac{365 \text{ days}}{1 \text{ year}}$$

Step 4 Multiply the old measurement by the conversion factor:

$$14 \text{ years} \times \frac{365 \text{ days}}{1 \text{ year}} = 5{,}110 \text{ days}$$

The person's age is 5,110 days.

Practice Problem A cat has a mass of 2.31 kg. If there are 1,000 g in 1 kg, what is the mass of the cat in grams?

Using Fractions

A **fraction** is a number that compares a part to the whole. For example, in the fraction $\frac{2}{3}$, the 2 represents the part and the 3 represents the whole. In the fraction $\frac{2}{3}$, the top number, 2, is called the numerator. The bottom number, 3, is called the denominator.

Sometimes fractions are not written in their simplest form. To determine a fraction's **simplest form,** you must find the greatest common factor (GCF) of the numerator and denominator. The greatest common factor is the largest common factor of all the factors the two numbers have in common.

For example, because the number 3 divides into 12 and 30 evenly, it is a common factor of 12 and 30. However, because the number 6 is the largest number that evenly divides into 12 and 30, it is the **greatest common factor.**

After you find the greatest common factor, you can write a fraction in its simplest form. Divide both the numerator and the denominator by the greatest common factor. The number that results is the fraction in its **simplest form.**

Example Twelve of the 20 corn plants in a field are more than 1.5 m tall. What fraction of the corn plants in the field is 1.5 m tall?

Step 1 Write the fraction.

$$\frac{\text{part}}{\text{whole}} = \frac{12}{20}$$

Step 2 To find the GCF of the numerator and denominator, list all of the factors of each number.

Factors of 12: 1, 2, 3, 4, 6, 12 (the numbers that divide evenly into 12)

Factors of 20: 1, 2, 4, 5, 10, 20 (the numbers that divide evenly into 20)

Step 3 List the common factors.

1, 2, 4.

Step 4 Choose the greatest factor in the list of common factors.

The GCF of 12 and 20 is 4.

Step 5 Divide the numerator and denominator by the GCF.

$$\frac{12 \div 4}{20 \div 4} = \frac{3}{5}$$

In the field, $\frac{3}{5}$ of the corn plants are more than 1.5 m tall.

Practice Problem There are 90 duck eggs in a population. Of those eggs, 66 hatch over a one-week period. What fraction of the eggs hatch over a one-week period? Write the fraction in simplest form.

Math Skill Handbook

Calculating Ratios

A **ratio** is a comparison of two numbers by division.

Ratios can be written 3 to 5 or 3:5. Ratios also can be written as fractions, such as $\frac{3}{5}$. Ratios, like fractions, can be written in simplest form. Recall that a fraction is in **simplest form** when the greatest common factor (GCF) of the numerator and denominator is 1.

Example From a package of sunflower seeds, 40 seeds germinated and 64 did not. What is the ratio of germinated to not germinated seeds as a fraction in simplest form?

Step 1 Write the ratio as a fraction.

$$\frac{\text{germinated}}{\text{not germinated}} = \frac{40}{64}$$

Step 2 Express the fraction in simplest form. The GCF of 40 and 64 is 8.

$$\frac{40}{64} = \frac{40 \div 8}{64 \div 8} = \frac{5}{8}$$

The ratio of germinated to not germinated seeds is $\frac{5}{8}$.

Practice Problem Two children measure 100 cm and 144 cm in height. What is the ratio of their heights in simplest fraction form?

Using Decimals

A **decimal** is a fraction with a denominator of 10, 100, 1,000, or another power of 10. For example, 0.854 is the same as the fraction $\frac{854}{1,000}$.

In a decimal, the decimal point separates the ones place and the tenths place. For example, 0.27 means twenty-seven hundredths, or $\frac{27}{100}$, where 27 is the **number of units** out of 100 units. Any fraction can be written as a decimal using division.

Example Write $\frac{5}{8}$ as a decimal.

Step 1 Write a division problem with the numerator, 5, as the dividend and the denominator, 8, as the divisor. Write 5 as 5.000.

Step 2 Solve the problem.

$$\begin{array}{r} 0.625 \\ 8\overline{)5.000} \\ \underline{48} \\ 20 \\ \underline{16} \\ 40 \\ \underline{40} \\ 0 \end{array}$$

Therefore, $\frac{5}{8} = 0.625$.

Practice Problem Write $\frac{19}{25}$ as a decimal.

Using Percentages

The word *percent* means "out of one hundred." A **percent** is a ratio that compares a number to 100. Suppose you read that 77 percent of all fish on Earth live in the Pacific Ocean. That is the same as reading that the ratio of Earth's fish that live in the Pacific Ocean is $\frac{77}{100}$. To express a fraction as a percent, first find an equivalent decimal for the fraction. Then, multiply the decimal by 100 and add the percent symbol. For example, $\frac{1}{2} = 1 \div 2 = 0.5$. Then $0.5 \cdot 100 = 50 = 50\%$.

Example Express $\frac{13}{20}$ as a percent.

Step 1 Find the equivalent decimal for the fraction.

$$
\begin{array}{r}
0.65 \\
20\overline{)13.00} \\
\underline{120} \\
100 \\
\underline{100} \\
0
\end{array}
$$

Step 2 Rewrite the fraction $\frac{13}{20}$ as 0.65.

Step 3 Multiply 0.65 by 100 and add the % sign.

$0.65 \cdot 100 = 65 = 65\%$

So, $\frac{13}{20} = 65\%$.

Practice Problem In an experimental population of 365 sheep, 73 were brown. What percent of the sheep were brown?

Using Precision and Significant Digits

When you make a **measurement,** the value you record depends on the precision of the measuring instrument. When adding or subtracting numbers with different precision, the answer is rounded to the smallest number of decimal places of any number in the sum or difference. When multiplying or dividing, the answer is rounded to the smallest number of significant figures of any number being multiplied or divided. When counting the number of **significant figures,** all digits are counted except zeros at the end of a number with no decimal such as 2,500, and zeros at the beginning of a decimal such as 0.03020.

Example The lengths 5.28 and 5.2 are measured in meters. Find the sum of these lengths and report the sum using the least precise measurement.

Step 1 Find the sum.

$$
\begin{array}{rl}
5.28 \text{ m} & \text{2 digits after the decimal} \\
\underline{+\ 5.2\ \ \text{m}} & \text{1 digit after the decimal} \\
10.48 \text{ m} &
\end{array}
$$

Step 2 Round to one digit after the decimal because the least number of digits after the decimal of the numbers being added is 1.

The sum is 10.5 m.

Practice Problem Multiply the numbers in the example using the rule for multiplying and dividing. Report the answer with the correct number of significant figures.

Solving One-Step Equations

An **equation** is a statement that two things are equal. For example, $A = B$ is an equation that states that A is equal to B.

Sometimes one side of the equation will contain a **variable** whose value is not known. In the equation $3x = 12$, the variable is x.

The equation is solved when the variable is replaced with a value that makes both sides of the equation equal to each other. For example, the solution of the equation $3x = 12$ is $x = 4$. If the x is replaced with 4, then the equation becomes $3 \cdot 4 = 12$, or $12 = 12$.

To solve an equation such as $8x = 40$, divide both sides of the equation by the number that multiplies the variable.

$$8x = 40$$
$$\frac{8x}{8} = \frac{40}{8}$$
$$x = 5$$

You can check your answer by replacing the variable with your solution and seeing if both sides of the equation are the same.

$$8x = 8 \cdot 5 = 40$$

The left and right sides of the equation are the same, so $x = 5$ is the solution.

Sometimes an equation is written in this way: $a = bc$. This also is called a **formula.** The letters can be replaced by numbers, but the numbers must still make both sides of the equation the same.

Example 1 Solve the equation $10x = 35$.

Step 1 Find the solution by dividing each side of the equation by 10.

$$10x = 35 \qquad \frac{10x}{10} = \frac{35}{10} \qquad x = 3.5$$

Step 2 Check the solution.

$$10x = 35 \qquad 10 \times 3.5 = 35 \qquad 35 = 35$$

Both sides of the equation are equal, so $x = 3.5$ is the solution to the equation.

Example 2 In the formula $a = bc$, find the value of c if $a = 20$ and $b = 2$.

Step 1 Rearrange the formula so the unknown value is by itself on one side of the equation by dividing both sides by b.

$$a = bc$$
$$\frac{a}{b} = \frac{bc}{b}$$
$$\frac{a}{b} = c$$

Step 2 Replace the variables a and b with the values that are given.

$$\frac{a}{b} = c$$
$$\frac{20}{2} = c$$
$$10 = c$$

Step 3 Check the solution.

$$a = bc$$
$$20 = 2 \times 10$$
$$20 = 20$$

Both sides of the equation are equal, so $c = 10$ is the solution when $a = 20$ and $b = 2$.

Practice Problem In the formula $h = gd$, find the value of d if $g = 12.3$ and $h = 17.4$.

Using Proportions

A **proportion** is an equation that shows that two ratios are equivalent. The ratios $\frac{2}{4}$ and $\frac{5}{10}$ are equivalent, so they can be written as $\frac{2}{4} = \frac{5}{10}$. This equation is an example of a proportion.

When two ratios form a proportion, the **cross products** are equal. To find the cross products in the proportion $\frac{2}{4} = \frac{5}{10}$, multiply the 2 and the 10, and the 4 and the 5. Therefore $2 \cdot 10 = 4 \cdot 5$, or $20 = 20$.

Because you know that both proportions are equal, you can use cross products to find a missing term in a proportion. This is known as **solving the proportion.** Solving a proportion is similar to solving an equation.

Example The heights of a tree and a pole are proportional to the lengths of their shadows. The tree casts a shadow of 24 m at the same time that a 6-m pole casts a shadow of 4 m. What is the height of the tree?

Step 1 Write a proportion.

$$\frac{\text{height of tree}}{\text{height of pole}} = \frac{\text{length of tree's shadow}}{\text{length of pole's shadow}}$$

Step 2 Substitute the known values into the proportion. Let h represent the unknown value, the height of the tree.

$$\frac{h}{6} = \frac{24}{4}$$

Step 3 Find the cross products.

$$h \cdot 4 = 6 \cdot 24$$

Step 4 Simplify the equation.

$$4h = 144$$

Step 5 Divide each side by 4.

$$\frac{4h}{4} = \frac{144}{4}$$

$$h = 36$$

The height of the tree is 36 m.

Practice Problem The proportions of bluefish are stable by the time they reach a length of 30 cm. The distance from the tip of the mouth to the back edge of the gill cover in a 35-cm bluefish is 15 cm. What is the distance from the tip of the mouth to the back edge of the gill cover in a 59-cm bluefish?

Using Statistics

Statistics is the branch of mathematics that deals with collecting, analyzing, and presenting data. In statistics, there are three common ways to summarize the data with a single number—the mean, the median, and the mode.

The **mean** of a set of data is the arithmetic average. It is found by adding the numbers in the data set and dividing by the number of items in the set.

The **median** is the middle number in a set of data when the data are arranged in numerical order. If there were an even number of data points, the median would be the mean of the two middle numbers.

The **mode** of a set of data is the number or item that appears most often.

Another number that often is used to describe a set of data is the range. The **range** is the difference between the largest number and the smallest number in a set of data.

A **frequency table** shows how many times each piece of data occurs, usually in a survey. The frequency table below shows the results of a student survey on favorite color.

Color	Tally	Frequency
red	IIII	4
blue	HHI	5
black	II	2
green	III	3
purple	HHI II	7
yellow	HHI I	6

Based on the frequency table data, which color is the favorite?

Example The high temperatures (in °C) on five consecutive days in a desert habitat under study are 39°, 37°, 44°, 36°, and 44°. Find the mean, median, mode, and range of this set.

To find the mean:
Step 1 Find the sum of the numbers.

$$39 + 37 + 44 + 36 + 44 = 200$$

Step 2 Divide the sum by the number of items, which is 5.

$$200 \div 5 = 40$$

The mean high temperature is 40°C.

To find the median:
Step 1 Arrange the temperatures from least to greatest.

$$36, \ 37, \ \underline{39}, \ 44, \ 44$$

Step 2 Determine the middle temperature.

The median high temperature is 39°C.

To find the mode:
Step 1 Group the numbers that are the same together.

44, 44, 36, 37, 39

Step 2 Determine the number that occurs most in the set.

<u>44, 44</u>, 36, 37, 39

The mode measure is 44°C.

To find the range:
Step 1 Arrange the temperatures from largest to smallest.

44, 44, 39, 37, 36

Step 2 Determine the largest and smallest temperature in the set.

<u>44</u>, 44, 39, 37, <u>36</u>

Step 3 Find the difference between the largest and smallest temperatures.

$$44 - 36 = 8$$

The range is 8°C.

Practice Problem Find the mean, median, mode, and range for the data set 8, 4, 12, 8, 11, 14, 16.

Safety in the Science Classroom

1. Always obtain your teacher's permission to begin an investigation.

2. Study the procedure. If you have questions, ask your teacher. Be sure you understand any safety symbols shown on the page.

3. Use the safety equipment provided for you. Goggles and a safety apron should be worn during most investigations.

4. Always slant test tubes away from yourself and others when heating them or adding substances to them.

5. Never eat or drink in the lab, and never use lab glassware as food or drink containers. Never inhale chemicals. Do not taste any substances or draw any material into a tube with your mouth.

6. Report any spill, accident, or injury, no matter how small, immediately to your teacher, then follow his or her instructions.

7. Know the location and proper use of the fire extinguisher, safety shower, fire blanket, first aid kit, and fire alarm.

8. Keep all materials away from open flames. Tie back long hair and tie down loose clothing.

9. If your clothing should catch fire, smother it with the fire blanket, or get under a safety shower. NEVER RUN.

10. If a fire should occur, turn off the gas; then leave the room according to established procedures.

Follow these procedures as you clean up your work area

1. Turn off the water and gas. Disconnect electrical devices.

2. Clean all pieces of equipment and return all materials to their proper places.

3. Dispose of chemicals and other materials as directed by your teacher. Place broken glass and solid substances in the proper containers. Make sure never to discard materials in the sink.

4. Clean your work area. Wash your hands thoroughly after working in the laboratory.

First Aid	
Injury	**Safe Response** ALWAYS NOTIFY YOUR TEACHER IMMEDIATELY
Burns	Apply cold water.
Cuts and Bruises	Stop any bleeding by applying direct pressure. Cover cuts with a clean dressing. Apply ice packs or cold compresses to bruises.
Fainting	Leave the person lying down. Loosen any tight clothing and keep crowds away.
Foreign Matter in Eye	Flush with plenty of water. Use eyewash bottle or fountain.
Poisoning	Note the suspected poisoning agent.
Any Spills on Skin	Flush with large amounts of water or use safety shower.

Care and Use of a Microscope

Eyepiece Contains magnifying lenses you look through.

Arm Supports the body tube.

Low-power objective Contains the lens with the lowest power magnification.

Stage clips Hold the microscope slide in place.

Fine adjustment Sharpens the image under high magnification.

Coarse adjustment Focuses the image under low power.

Body tube Connects the eyepiece to the revolving nosepiece.

Revolving nosepiece Holds and turns the objectives into viewing position.

High-power objective Contains the lens with the highest magnification.

Stage Supports the microscope slide.

Light source Provides light that passes upward through the diaphragm, the specimen, and the lenses.

Base Provides support for the microscope.

Caring for a Microscope

1. Always carry the microscope holding the arm with one hand and supporting the base with the other hand.

2. Don't touch the lenses with your fingers.

3. The coarse adjustment knob is used only when looking through the lowest-power objective lens. The fine adjustment knob is used when the high-power objective is in place.

4. Cover the microscope when you store it.

Using a Microscope

1. Place the microscope on a flat surface that is clear of objects. The arm should be toward you.

2. Look through the eyepiece. Adjust the diaphragm so light comes through the opening in the stage.

3. Place a slide on the stage so the specimen is in the field of view. Hold it firmly in place by using the stage clips.

4. Always focus with the coarse adjustment and the low-power objective lens first. After the object is in focus on low power, turn the nosepiece until the high-power objective is in place. Use ONLY the fine adjustment to focus with the high-power objective lens.

Making a Wet-Mount Slide

1. Carefully place the item you want to look at in the center of a clean, glass slide. Make sure the sample is thin enough for light to pass through.

2. Use a dropper to place one or two drops of water on the sample.

3. Hold a clean coverslip by the edges and place it at one edge of the water. Slowly lower the coverslip onto the water until it lies flat.

4. If you have too much water or a lot of air bubbles, touch the edge of a paper towel to the edge of the coverslip to draw off extra water and draw out unwanted air.

Diversity of Life: Classification of Living Organisms

A six-kingdom system of classification of organisms is used today. Two kingdoms—Kingdom Archaebacteria and Kingdom Eubacteria—contain organisms that do not have a nucleus and that lack membrane-bound structures in the cytoplasm of their cells. The members of the other four kingdoms have a cell or cells that contain a nucleus and structures in the cytoplasm, some of which are surrounded by membranes. These kingdoms are Kingdom Protista, Kingdom Fungi, Kingdom Plantae, and Kingdom Animalia.

Kingdom Archaebacteria

one-celled; some absorb food from their surroundings; some are photosynthetic; some are chemosynthetic; many are found in extremely harsh environments including salt ponds, hot springs, swamps, and deep-sea hydrothermal vents

Kingdom Eubacteria

one-celled; most absorb food from their surroundings; some are photosynthetic; some are chemosynthetic; many are parasites; many are round, spiral, or rod-shaped; some form colonies

Kingdom Protista

Phylum Euglenophyta one-celled; photosynthetic or take in food; most have one flagellum; euglenoids

Phylum Bacillariophyta one-celled; photosynthetic; have unique double shells made of silica; diatoms

Phylum Dinoflagellata one-celled; photosynthetic; contain red pigments; have two flagella; dinoflagellates

Phylum Chlorophyta one-celled, many-celled, or colonies; photosynthetic; contain chlorophyll; live on land, in freshwater, or salt water; green algae

Phylum Rhodophyta most are many-celled; photosynthetic; contain red pigments; most live in deep, saltwater environments; red algae

Phylum Phaeophyta most are many-celled; photosynthetic; contain brown pigments; most live in saltwater environments; brown algae

Phylum Rhizopoda one-celled; take in food; are free-living or parasitic; move by means of pseudopods; amoebas

Kingdom Eubacteria
Bacillus anthracis

Phylum Chlorophyta
Desmids

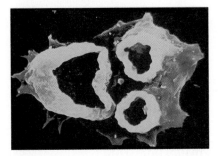

Amoeba

Phylum Zoomastigina one-celled; take in food; free-living or parasitic; have one or more flagella; zoomastigotes

Phylum Ciliophora one-celled; take in food; have large numbers of cilia; ciliates

Phylum Sporozoa one-celled; take in food; have no means of movement; are parasites in animals; sporozoans

Phylum Myxomycota
Slime mold

Phyla Myxomycota and Acrasiomycota one- or many-celled; absorb food; change form during life cycle; cellular and plasmodial slime molds

Phylum Oomycota many-celled; are either parasites or decomposers; live in freshwater or salt water; water molds, rusts and downy mildews

Kingdom Fungi

Phylum Zygomycota many-celled; absorb food; spores are produced in sporangia; zygote fungi; bread mold

Phylum Ascomycota one- and many-celled; absorb food; spores produced in asci; sac fungi; yeast

Phylum Basidiomycota many-celled; absorb food; spores produced in basidia; club fungi; mushrooms

Phylum Deuteromycota members with unknown reproductive structures; imperfect fungi; *Penicillium*

Mycophycota organisms formed by symbiotic relationship between an ascomycote or a basidiomycote and green alga or cyanobacterium; lichens

Phylum Oomycota
Phytophthora infestans

Lichens

Kingdom Plantae

Divisions Bryophyta (mosses), **Anthocerophyta** (hornworts), **Hepatophytal** (liverworts), **Psilophytal** (whisk ferns) many-celled nonvascular plants; reproduce by spores produced in capsules; green; grow in moist, land environments

Division Lycophyta many-celled vascular plants; spores are produced in conelike structures; live on land; are photosynthetic; club mosses

Division Sphenophyta vascular plants; ribbed and jointed stems; scalelike leaves; spores produced in conelike structures; horsetails

Division Pterophyta vascular plants; leaves called fronds; spores produced in clusters of sporangia called sori; live on land or in water; ferns

Division Ginkgophyta deciduous trees; only one living species; have fan-shaped leaves with branching veins and fleshy cones with seeds; ginkgoes

Division Cycadophyta palmlike plants; have large, featherlike leaves; produces seeds in cones; cycads

Division Coniferophyta deciduous or evergreen; trees or shrubs; have needlelike or scalelike leaves; seeds produced in cones; conifers

Division Gnetophyta shrubs or woody vines; seeds are produced in cones; division contains only three genera; gnetum

Division Anthophyta dominant group of plants; flowering plants; have fruits with seeds

Kingdom Animalia

Phylum Porifera aquatic organisms that lack true tissues and organs; are asymmetrical and sessile; sponges

Phylum Cnidaria radially symmetrical organisms; have a digestive cavity with one opening; most have tentacles armed with stinging cells; live in aquatic environments singly or in colonies; includes jellyfish, corals, hydra, and sea anemones

Phylum Platyhelminthes bilaterally symmetrical worms; have flattened bodies; digestive system has one opening; parasitic and free-living species; flatworms

Division Anthophyta
Tomato plant

Division Bryophyta
Liverwort

Phylum Platyhelminthes
Flatworm

Phylum Chordata

Phylum Nematoda round, bilaterally symmetrical body; have digestive system with two openings; free-living forms and parasitic forms; roundworms

Phylum Mollusca soft-bodied animals, many with a hard shell and soft foot or footlike appendage; a mantle covers the soft body; aquatic and terrestrial species; includes clams, snails, squid, and octopuses

Phylum Annelida bilaterally symmetrical worms; have round, segmented bodies; terrestrial and aquatic species; includes earthworms, leeches, and marine polychaetes

Phylum Arthropoda largest animal group; have hard exoskeletons, segmented bodies, and pairs of jointed appendages; land and aquatic species; includes insects, crustaceans, and spiders

Phylum Echinodermata marine organisms; have spiny or leathery skin and a water-vascular system with tube feet; are radially symmetrical; includes sea stars, sand dollars, and sea urchins

Phylum Chordata organisms with internal skeletons and specialized body systems; most have paired appendages; all at some time have a notochord, nerve cord, gill slits, and a postanal tail; include fish, amphibians, reptiles, birds, and mammals

This glossary defines each key term that appears in bold type in the text. It also shows the chapter, section, and page number where you can find the words used.

A

abiotic: nonliving, physical features of the environment, including air, water, sunlight, soil, temperature, and climate. (Chap. 2, Sec. 1, p. 36)

acid precipitation: precipitation with a pH below 5.6—which occurs when air pollutants from the burning of fossil fuels react with water in the atmosphere to form strong acids—that can pollute water, kill fish and plants, and damage soils. (Chap. 4, Sec. 2, p. 103)

acid rain: forms when sulfur dioxide and nitrogen oxide from industries and car exhausts combine with water vapor in the air; can wash nutrients from soil and damage trees and aquatic life. (Chap. 5, Sec. 1, p. 135)

atmosphere: air surrounding Earth; is made up of gases, including 78 percent nitrogen, 21 percent oxygen, and 0.03 percent carbon dioxide. (Chap. 2, Sec. 1, p. 37)

B

biodiversity: variety of life in an ecosystem, most commonly measured by the number of species that live in a given area. (Chap. 5, Sec. 1, p. 126)

biomes (BI ohmz): large geographic areas with similar climates and ecosystems; includes tundra, taiga, desert, temperate deciduous forest, temperate rain forest, tropical rain forest, and grassland. (Chap. 3, Sec. 2, p. 68)

biosphere: part of Earth that supports life, including the top portion of Earth's crust, the atmosphere, and all the water on Earth's surface. (Chap. 1, Sec. 1, p. 8)

biotic (bi AHT ik): features of the environment that are alive or were once alive. (Chap. 2, Sec. 1, p. 36)

C

captive population: population of organisms that is cared for by humans. (Chap. 5, Sec. 2, p. 142)

carbon cycle: model describing how carbon molecules move between the living and nonliving world. (Chap. 2, Sec. 2, p. 49)

carrying capacity: largest number of individuals of a particular species that an ecosystem can support over time. (Chap. 1, Sec. 2, p. 15)

chemosynthesis (kee moh SIN thuh sus): process in which producers make energy-rich nutrient molecules from chemicals. (Chap. 2, Sec. 3, p. 51)

climate: average weather conditions of an area over time, including wind, temperature, and rainfall or other types of precipitation such as snow or sleet. (Chap. 2, Sec. 1, p. 41)

climax community: stable, end stage of ecological succession in which the plants and animals of a community use resources efficiently and balance is maintained by disturbances such as fire. (Chap. 3, Sec. 1, p. 67)

commensalism: a type of symbiotic relationship in which one organism benefits and the other organism is not affected. (Chap. 1, Sec. 3, p. 22)

community: all the populations of different species that live in an ecosystem. (Chap. 1, Sec. 1, p. 10)

condensation: process that takes place when a gas changes to a liquid. (Chap. 2, Sec. 2, p. 45)

conservation biology: study of methods for protecting Earth's biodiversity; uses strategies such as reintroduction programs and habitat restoration and works to preserve threatened and endangered species. (Chap. 5, Sec. 2, p. 138)

English Glossary

consumer: organism that cannot create energy-rich molecules but obtains its food by eating other organisms. (Chap. 1, Sec. 3, p. 21)

coral reef: diverse ecosystem formed from the calcium carbonate shells secreted by corals. (Chap. 3, Sec. 3, p. 81)

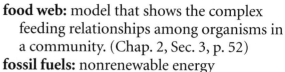

D

desert: driest biome on Earth with less than 25 cm of rain each year; has dunes or thin soil with little organic matter and plants and animals specially adapted to survive extreme conditions. (Chap. 3, Sec. 2, p. 74)

E

ecology: study of the interactions that take place among organisms and their environment. (Chap. 1, Sec. 1, p. 9)

ecosystem: all the living organisms that live in an area and the nonliving features of their environment. (Chap. 1, Sec. 1, p. 9)

endangered species: species that is in danger of becoming extinct because of factors such as habitat loss, overhunting, and pollution. (Chap. 5, Sec. 1, p. 131)

energy pyramid: model that shows the amount of energy available at each feeding level in an ecosystem. (Chap. 2, Sec. 3, p. 53)

erosion: movement of soil from one place to another. (Chap. 4, Sec. 2, p. 109)

estuary: extremely fertile area where a river meets an ocean; contains a mixture of freshwater and saltwater and serves as a nursery for many species of fish. (Chap. 3, Sec. 3, p. 82)

evaporation: process that takes place when a liquid changes to a gas. (Chap. 2, Sec. 2, p. 44)

extinct species: species that was once present on Earth but has died out. (Chap. 5, Sec. 1, p. 130)

F

food web: model that shows the complex feeding relationships among organisms in a community. (Chap. 2, Sec. 3, p. 52)

fossil fuels: nonrenewable energy resources—coal, oil, and natural gas—that formed in Earth's crust over hundreds of millions of years. (Chap. 4, Sec. 1, p. 96)

G

geothermal energy: heat energy within Earth's crust, available only where natural geysers or volcanoes are located. (Chap. 4, Sec. 1, p. 99)

grasslands: temperate and tropical regions with 25 cm to 75 cm of precipitation each year that are dominated by climax communities of grasses; ideal for growing crops and raising cattle and sheep. (Chap. 3, Sec. 2, p. 75)

greenhouse effect: heat-trapping feature of the atmosphere that keeps Earth warm enough to support life. (Chap. 4, Sec. 2, p. 104)

H

habitat: place where an organism lives and that provides the types of food, shelter, moisture, and temperature needed for survival. (Chap. 1, Sec. 1, p. 11)

habitat restoration: process of bringing a damaged habitat back to a healthy condition. (Chap. 5, Sec. 2, p. 141)

hazardous wastes: waste materials, such as pesticides and leftover paints, that are harmful to human health or poisonous to living organisms. (Chap. 4, Sec. 2, p. 110)

hydroelectric power: electricity produced when the energy of falling water turns the blades of a generator turbine. (Chap. 4, Sec. 1, p. 97)

I

intertidal zone: part of the shoreline that is underwater at high tide and exposed to the air at low tide. (Chap. 3, Sec. 3, p. 82)

introduced species: species that moves into an ecosystem as a result of human actions and can reduce or eliminate populations of native species. (Chap. 5, Sec. 1, p. 134)

L

limiting factor: anything that can restrict the size of a population, including living and nonliving features of an ecosystem, such as predators or drought. (Chap. 1, Sec. 2, p. 14)

M

mutualism: a type of symbiotic relationship in which both organisms benefit. (Chap. 1, Sec. 3, p. 22)

N

native species: original organisms in an ecosystem. (Chap. 5, Sec. 1, p. 134)

natural resources: parts of Earth's environment that supply materials useful or necessary for the survival of living organisms. (Chap. 4, Sec. 1, p. 94)

niche: in an ecosystem, refers to the unique ways an organism survives, obtains food and shelter, and avoids danger. (Chap. 1, Sec. 3, p. 23)

nitrogen cycle: model describing how nitrogen moves from the atmosphere to the soil, to living organisms, and then back to the atmosphere. (Chap. 2, Sec. 2, p. 46)

nitrogen fixation: process in which some types of bacteria in the soil change nitrogen gas into a form of nitrogen that plants can use. (Chap. 2, Sec. 2, p. 46)

nonrenewable resources: natural resources, such as petroleum, minerals, and metals, that are used more quickly than they can be replaced by natural processes. (Chap. 4, Sec. 1, p. 95)

nuclear energy: energy produced from the splitting apart of billions of uranium nuclei by a nuclear fission reaction. (Chap. 4, Sec. 1, p. 98)

O

ozone depletion: thinning of Earth's ozone layer caused by chlorofluorocarbons (CFCs) leaking into the air and reacting chemically with ozone, breaking the ozone molecules apart. (Chap. 4, Sec. 2, p. 105)

ozone layer: protective layer of gas in Earth's atmosphere, about 15 km to 30 km above Earth's surface, that absorbs ultraviolet radiation from the Sun. (Chap. 5, Sec. 1, p. 136)

P

parasitism: a type of symbiotic relationship in which one organism benefits and the other organism is harmed. (Chap. 1, Sec. 3, p. 22)

petroleum: nonrenewable resource formed over hundreds of millions of years mostly from the remains of microscopic marine organisms buried in Earth's crust. (Chap. 4, Sec. 1, p. 95)

pioneer species: a group of hardy organisms, such as lichens, found in the primary stage of succession and that begin an area's soil-building process. (Chap. 3, Sec. 1, p. 64)

pollutant: substance that contaminates any part of the environment. (Chap. 4, Sec. 2, p. 102)

population: all the organisms that belong to the same species living in a community. (Chap. 1, Sec. 1, p. 10)

English Glossary

producer: organism, such as a green plant or alga, that uses an outside source of energy like the Sun to create energy-rich food molecules. (Chap. 1, Sec. 3, p. 20)

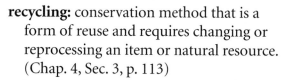

recycling: conservation method that is a form of reuse and requires changing or reprocessing an item or natural resource. (Chap. 4, Sec. 3, p. 113)

reintroduction program: conservation strategy that returns organisms to an area where the species once lived and may involve seed banks, captive populations, and relocation. (Chap. 5, Sec. 2, p. 142)

renewable resources: natural resources, such as water, sunlight, and crops, that are constantly being recycled or replaced by nature. (Chap. 4, Sec. 1, p. 94)

S

soil: mixture of mineral and rock particles, the remains of dead organisms, air, and water that forms the topmost layer of Earth's crust and supports plant growth. (Chap. 2, Sec. 1, p. 38)

succession: natural, gradual changes in the types of species that live in an area; can be primary or secondary. (Chap. 3, Sec. 1, p. 64)

symbiosis: any close relationship between species, including mutualism, commensalism, and parasitism. (Chap. 1, Sec. 3, p. 22)

T

taiga (TI guh): world's largest biome, located south of the tundra between 50°N and 60°N latitude; has long, cold winters, precipitation between 35 cm and 100 cm each year, cone-bearing evergreen trees, and dense forests. (Chap. 3, Sec. 2, p. 70)

temperate deciduous forest: biome usually having four distinct seasons, annual precipitation between 75 cm and 150 cm, and climax communities of deciduous trees. (Chap. 3, Sec. 2, p. 71)

temperate rain forest: biome with 200 cm to 400 cm of precipitation each year, average temperatures between 9°C and 12°C, and forests dominated by trees with needlelike leaves. (Chap. 3, Sec. 2, p. 71)

threatened species: species that is likely to become endangered in the near future because of factors such as overhunting, introduced species, and pollution. (Chap. 5, Sec. 1, p. 131)

tropical rain forest: most biologically diverse biome; has an average temperature of 25°C and receives between 200 cm and 600 cm of precipitation each year. (Chap. 3, Sec. 2, p. 72)

tundra: cold, dry, treeless biome with less than 25 cm of precipitation each year, a short growing season, permafrost, and winters that can be six to nine months long. (Chap. 3, Sec. 2, p. 69)

water cycle: model describing how water moves from Earth's surface to the atmosphere and back to the surface again through evaporation, condensation, and precipitation. (Chap. 2, Sec. 2, p. 45)

wetland: a region that is wet most or all of the year. (Chap. 3, Sec. 3, p. 79)

Este glosario define cada término clave que aparece en negrillas en el texto. También muestra el capítulo, la sección y el número de página en donde se usa dicho término.

A

abiotic / abióticos: factores físicos inanimados del medio ambiente, que incluyen el aire, el agua, la luz solar, el suelo, la temperatura y el clima. (Cap. 2, Sec. 1, pág. 36)

acid precipitation / precipitación ácida: precipitación con un pH menor de 5.6 (la cual ocurre cuando los contaminantes del aire, provenientes de la quema de combustibles fósiles, reaccionan con el agua de la atmósfera para formar ácidos fuertes) capaz de contaminar el agua, matar peces y plantas y perjudicar los suelos. (Cap. 4, Sec. 2, pág. 103)

acid rain / lluvia ácida: se produce cuando el dióxido de azufre y el óxido de nitrógeno provenientes de las industrias y de los tubos de escape de los vehículos se combinan con el vapor de agua en el aire; puede arrastrar nutrientes del suelo y perjudicar los árboles y la vida acuática. (Cap. 5, Sec. 1, pág. 135)

atmosphere / atmósfera: el aire que rodea la Tierra; está compuesta por gases, entre los cuales se incluye un 78 por ciento de nitrógeno, un 21 por ciento de oxígeno y 0.03 por ciento de dióxido de carbono. (Cap. 2, Sec. 1, pág. 37)

B

biodiversity / biodiversidad: variedad de vida en un ecosistema, que se mide principalmente por el número de especies que viven en un área dada. (Cap. 5, Sec. 1, pág. 126)

biomes / biomas: áreas geográficas extensas con climas y ecosistemas similares; incluye la tundra, la taiga, el desierto, los bosques caducifolios de zonas templadas, los bosques pluviales de zonas templadas, los bosques pluviales tropicales y las praderas. (Cap. 3, Sec. 2, pág. 68)

biosphere / biosfera: parte de la Tierra que sostiene la vida; incluye la parte superior de la corteza terrestre, la atmósfera y toda el agua sobre la superficie de la Tierra. (Cap. 1, Sec. 1, pág. 8)

biotic / bióticos: factores del medio ambiente que son seres vivos o seres que alguna vez estuvieron vivos. (Cap. 2, Sec. 1, pág. 36)

C

captive population / población en cautiverio: población de organismos que los seres humanos cuidan. (Cap. 5, Sec. 2, pág. 142)

carbon cycle / ciclo del carbono: modelo que describe cómo las células del carbono se movilizan entre el mundo vivo y el inanimado. (Cap. 2, Sec. 2, pág. 49)

carrying capacity / capacidad de carga: el número mayor de individuos de una especie en particular que puede mantener un ecosistema de manera prolongada. (Cap. 1, Sec. 2, pág. 15)

chemosynthesis / quimiosíntesis: proceso en el cual los productores elaboran moléculas nutritivas ricas en energía a partir de sustancias químicas. (Cap. 2, Sec. 3, pág. 51)

climate / clima: condiciones meteorológicas promedio de una región durante un período de tiempo, entre las cuales se incluyen el viento, la temperatura y la precipitación pluvial u otro tipo de precipitación como la nieve o la cellisca. (Cap. 2, Sec. 1, pág. 41)

climax community / comunidad clímax: etapa final estable de la sucesión ecológica, en que las plantas y los animales de una comunidad usan eficientemente los recursos y en la cual se mantiene el equilibrio mediante perturbaciones como los incendios. (Cap. 3, Sec. 1, pág. 67)

commensalism / comensalismo: tipo de relación simbiótica en el cual un organismo se beneficia y el otro organismo no se ve afectado. (Cap. 1, Sec. 3, pág. 22)

community / comunidad: todas las poblaciones de diferentes especies que viven en un ecosistema. (Cap. 1, Sec. 1, pág. 10)

condensation / condensación: proceso que se efectúa cuando un gas se convierte en un líquido. (Cap. 2, Sec. 2, pág. 45)

conservation biology / biología de la conservación: estudio de métodos para proteger la biodiversidad terrestre; utiliza estrategias como programas de reintroducción y restauración de hábitats y se esfuerza por preservar las especies amenazadas y aquellas en peligro de extinción. (Cap. 5, Sec. 2, pág. 138)

consumer / consumidor: organismo que no puede fabricar moléculas ricas en energía, sino que obtiene su alimento al alimentarse de otros organismos. (Cap. 1, Sec. 3, pág. 21)

coral reef / arrecife de coral: ecosistema diverso formado de las conchas de carbonato de calcio secretadas por los corales. (Cap. 3, Sec. 3, pág. 81)

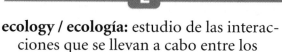

desert / desierto: el bioma más seco de la Tierra, con menos de 25 cm de lluvia anual; tiene dunas o suelo delgado con poca materia orgánica, y plantas y animales especialmente adaptados para sobrevivir condiciones extremas. (Cap. 3, Sec. 2, pág. 74)

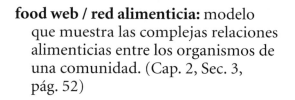

ecology / ecología: estudio de las interacciones que se llevan a cabo entre los organismos y su ambiente. (Cap. 1, Sec. 1, pág. 9)

ecosystem / ecosistema: todos los organismos vivos que habitan en un área y las cosas inanimadas en su ambiente. (Cap. 1, Sec. 1, pág. 9)

endangered species / especie en peligro de extinción: especie que está en peligro de extinguirse debido a factores como la pérdida de hábitat, la caza excesiva y la contaminación. (Cap. 5, Sec. 1, pág. 131)

energy pyramid / pirámide de energía: modelo que muestra la cantidad de energía disponible en cada nivel alimenticio de un ecosistema. (Cap. 2, Sec. 3, pág. 53)

erosion / erosión: movimiento del suelo de un lugar a otro. (Cap. 4, Sec. 2, pág. 109)

estuary / estuario: área estremadamente fértil donde un río desemboca en un océano; contiene una mezcla de agua dulce y agua salada y sirve de vivero para muchas especies de peces. (Cap. 3, Sec. 3, pág. 82)

evaporation / evaporación: proceso que se lleva a cabo cuando un líquido se convierte en un gas. (Cap. 2, Sec. 2, pág. 44)

extinct species / especie extinta: especie que alguna vez vivió en la Tierra, pero que ha desaparecido. (Cap. 5, Sec. 1, pág. 130)

food web / red alimenticia: modelo que muestra las complejas relaciones alimenticias entre los organismos de una comunidad. (Cap. 2, Sec. 3, pág. 52)

fossil fuels / combustibles fósiles: recursos energéticos no renovables (carbón, petróleo y gas natural) que se formaron en la corteza terrestre durante un período de cientos de millones de años. (Cap. 4, Sec. 1, pág. 96)

geothermal energy / energía geotérmica: energía calorífica en el interior de la corteza terrestre, disponible solamente cuando se localizan géisers naturales o volcanes. (Cap. 4, Sec. 1, pág. 99)

grasslands / praderas: regiones tropicales y de zonas templadas cuya precipitación anual varía de 25 a 75 cm; dominadas por comunidades clímax de pastos; ideales para el crecimiento de cosechas y el pastoreo del ganado vacuno y bovino. (Cap. 3, Sec. 2, pág. 75)

greenhouse effect / efecto invernadero: fenómeno de la atmósfera que atrapa el calor y que mantiene la Tierra con suficiente calor como para sostener la vida. (Cap. 4, Sec. 2, pág. 104)

H

habitat / hábitat: lugar en donde vive un organismo y que le provee los tipos de alimento, refugio, humedad y temperaturas necesarias para la sobrevivencia. (Cap. 1, Sec. 1, pág. 11)

habitat restoration / restauración de hábitats: proceso de restablecer condiciones saludables en un hábitat que ha sufrido daño. (Cap. 5, Sec. 2, pág. 141)

hazardous wastes / residuos peligrosos: materiales de desecho, como los pesticidas y los restos de pintura, que son dañinos para la salud humana o venenosos para los organismos vivos. (Cap. 4, Sec. 2, pág. 110)

hydroelectric power / energía hidroeléctrica: electricidad producida cuando la fuerza hidráulica hace girar las aspas de una turbina generadora de electricidad. (Cap. 4, Sec. 1, pág. 97)

I

intertidal zone / zona intermareal: parte de la costa cubierta de agua durante la marea alta y que está expuesta al aire durante la marea baja. (Cap. 3, Sec. 3, pág. 82)

introduced species / especie introducida: especie que llega a un ecosistema como resultado de acciones humanas y que puede reducir o eliminar poblaciones de especies nativas. (Cap. 5, Sec. 1, pág. 134)

L

limiting factor / factor limitativo: cualquier cosa que puede limitar el tamaño de una población, incluye los rasgos vivos y los inanimados de un ecosistema, como los depredadores o la sequía. (Cap. 1, Sec. 2, pág. 14)

M

mutualism / mutualismo: tipo de relación simbiótica en que ambos organismos se benefician. (Cap. 1, Sec. 3, pág. 22)

N

native species / especie autóctona: organismos originales de un ecosistema. (Cap. 5, Sec. 1, pág. 134)

natural resources / recursos naturales: partes del medio ambiente terrestre que suministran materiales útiles o necesarios para la sobreviviencia de los organismos vivos. (Cap. 4, Sec. 1, pág. 94)

niche / nicho: en un ecosistema, se refiere a las maneras particulares en que un organismo sobrevive, obtiene alimentos y refugio y evita peligros. (Cap. 1, Sec. 3, pág. 23)

nitrogen cycle / ciclo del nitrógeno: modelo que describe cómo se mueve el nitrógeno de la atmósfera al suelo, pasando luego a los organismos vivos y, finalmente, de regreso a la atmósfera. (Cap. 2, Sec. 2, pág. 46)

nitrogen fixation / fijación del nitrógeno: proceso en el cual algunos tipos de bacterias que se hallan en el suelo convierten el gas de nitrógeno en una forma de nitrógeno que pueden usar las plantas. (Cap. 2, Sec. 2, pág. 46)

nonrenewable resources / recursos no renovables: recursos naturales, como el petróleo crudo, los minerales y los metales, que se usan con mayor rapidez de lo que se pueden reemplazar por procesos naturales. (Cap. 4, Sec. 1, pág. 95)

nuclear energy / energía nuclear: energía producida a partir de la separación de billones de núcleos de uranio mediante una reacción de fisión nuclear. (Cap. 4, Sec. 1, pág. 98)

O

ozone depletion / agotamiento del ozono: enrarecimiento de la capa de ozono de la Tierra debido a los clorofluorocarbonos que se escapan al aire y reaccionan químicamente con el ozono, separando sus moléculas. (Cap. 4, Sec. 2, pág. 105)

ozone layer / capa de ozono: capa de gas protectora en la atmósfera terrestre, unos 15 km a 30 km encima de la superficie de la Tierra, que absorbe la radiación ultravioleta proveniente del Sol. (Cap. 5, Sec. 1, pág. 136)

P

parasitism / parasitismo: tipo de relación simbiótica en que un organismo se beneficia y el otro organismo es perjudicado. (Cap. 1, Sec. 3, pág. 22)

petroleum / petróleo crudo: recurso no renovable que se ha formado durante el transcurso de cientos de millones de años, principalmente de los restos de organismos marinos microscópicos enterrados en la corteza terrestre. (Cap. 4, Sec. 1, pág. 95)

pioneer species / especie pionera: grupo de organismos, como los líquenes, que se encuentran en la etapa primaria de sucesión y que comienza el proceso de formación del suelo. (Cap. 3, Sec. 1, pág. 64)

pollutant / contaminante: sustancia que contamina cualquier parte del medio ambiente. (Cap. 4, Sec. 2, pág. 102)

population / población: todos los organismos que pertenecen a la misma especie y que viven en una comunidad. (Cap. 1, Sec. 1, pág. 10)

producer / productor: organismo que utiliza fuentes externas de energía como el Sol, para fabricar moléculas ricas en energía; por ejemplo, las plantas o algas verdes. (Cap. 1, Sec. 3, pág. 20)

R

recycling / reciclaje: método de conservación que constituye una forma de reutilización y que requiere la transformación o reprocesamiento de un artículo o de un recurso natural. (Cap. 4, Sec. 3, pág. 113)

reintroduction program / programa de reintroducción: estrategia de conservación que devuelve organismos a un área donde una vez vivieron tales especies y que puede implicar bancos de semilla, poblaciones en cautiverio y reubicación. (Cap. 5, Sec. 2, pág. 142)

renewable resources / recursos renovables: recursos naturales, como el agua, la luz solar y los cultivos, que la naturaleza recicla o reemplaza constantemente. (Cap. 4, Sec. 1, pág. 94)

S

soil / suelo: mezcla de partículas minerales y rocosas, restos de organismos muertos, aire y agua que forma la capa superior de la corteza terrestre y que sostiene el crecimiento vegetal. (Cap. 2, Sec. 1, pág. 38)

succession / sucesión: cambios graduales naturales en los tipos de especies que moran en un área; puede ser primaria o secundaria. (Cap. 3, Sec. 1, pág. 64)

symbiosis / simbiosis: cualquier relación estrecha entre especies, incluye el mutualismo, el comensalismo y el parasitismo. (Cap. 1, Sec. 3, pág. 22)

T

taiga / taiga: el bioma más grande del mundo, ubicado al sur de la tundra entre las latitudes de 50° N y 60° N; tiene inviernos fríos y largos, precipitación anual entre 35 cm y 100 cm, árboles coníferos siempreverdes y bosques densos. (Cap. 3, Sec. 2, pág. 70)

temperate deciduous forest / bosque caducifolio de zona templada: bioma que por lo general presenta cuatro estaciones bien marcadas, una precipitación anual entre 75 cm y 150 cm y comunidades clímax de árboles caducifolio. (Cap. 3, Sec. 2, pág. 71)

temperate rain forest / bosque pluvial de zona templada: bioma cuya precipitación anual varía de 200 cm a 400 cm, con temperaturas promedio entre 9°C y 12°C y bosques dominados por árboles con hojas en forma de agujas. (Cap. 3, Sec. 2, pág. 71)

threatened species / especie amenazada: especie propensa a hallarse en peligro de extinción en un futuro cercano, debido a factores como la caza excesiva, la introducción de especies y la contaminación. (Cap. 5, Sec. 1, pág. 131)

tropical rain forest / bosque pluvial tropical: el bioma más diverso biológicamente; tiene una temperatura promedio de 25°C y recibe entre 200 cm y 600 cm de precipitación anual. (Cap. 3, Sec. 2, pág. 72)

tundra / tundra: bioma sin árboles, frío y seco con menos de 25 cm de precipitación cada año, una temporada de crecimiento corta, permagel e inviernos que duran entre seis y nueve meses. (Cap. 3, Sec. 2, pág. 69)

W

water cycle / ciclo del agua: modelo que describe cómo se mueve el agua de la superficie de la Tierra a la atmósfera para regresar nuevamente a la superficie a través de la evaporación, condensación y precipitación. (Cap. 2, Sec. 2, pág. 45)

wetland / zonas pantanosas: áreas que se mantienen mojadas la mayor parte del año. (Cap. 3, Sec. 3, pág. 79)

Spanish Glossary

The index for *Ecology* will help you locate major topics in the book quickly and easily. Each entry in the index is followed by the number of the pages on which the entry is discussed. A page number given in boldfaced type indicates the page on which that entry is defined. A page number given in italic type indicates a page on which the entry is used in an illustration or photograph. The abbreviation *act.* indicates a page on which the entry is used in an activity.

A

Abiotic factors, 36–43, *36;* air, 37; climate, 41–42, *41, 42;* soil, 38, *act.* 43; sunlight, 38, *38;* temperature, 39–40, *39, 40;* water, 37, *37*

Acid precipitation, 103, *103,* **135,** *135*

Acid rain, 135, *135*

Activities, 25, 26–27, 43, 54–55, 76, 84–85, 111, 116–117, 137, 144–145

Agriculture: biodiversity and, 128, *128,* 129; nitrogen fixation and, 46, *46;* soil loss and, 109, *109*

Air: as abiotic factor in environment, 37

Air pollution, 102–106, 135, *135;* acid precipitation, 103, *103;* greenhouse effect, 104, *104, act.* 111; indoor, 106, *106;* ozone depletion, 105, *105;* smog, 102, *102*

Air temperature, 104, *104*

Algae: mutualism and, 22, *22;* as producers, 20; water pollution and, 107

Alligator(s), *24,* 139

Aluminum: recycling, 114

Animal(s): captive populations of, 142, *142;* competition among, 12, *12;* cooperation among, 24; endangered species of, 131, *131, 132,* 139, *139;* extinct species of, 130, *130;* food chain and, 51, *51;* habitats of, 11, *11,* 12, *12,* 23, *23,* 133–134, *133,* 140–141, *140, 141;* migration of, 17;

reintroduction programs for, 142; relocation of, 143; threatened species of, 131, *131, 132;* on tundra, 69, *69*

Appendices. *see* Reference Handbook

Aquatic ecosystems, 77–85; freshwater, 77–79, *77, 78, 79, act.* 84–85; saltwater, 80–83, *81, 82, 83*

Atmosphere, 37; as abiotic factor in environment, 37; gravity and, 41; ozone layer in, 136

B

Before You Read, 7, 35, 63, 93, 125

Biodiversity, 126–137, *126;* importance of, 127–129, *128, 129;* laws protecting, 139; measuring, 126–127, *127;* of plants, 128, *128, act.* 144–145; protecting, 138, *138;* reduction of, 130–136, *130, 131, 132*

Biological organization, 10, *10*

Biomes, 68. *see also* Land biomes

Biosphere, 8, *8*

Biotic factors, 36

Biotic potential, 16, 17

Birds: competition and, 12, *12;* extinct species of, 130, *130;* habitats of, 11, *11,* 12, *12;* interactions with other animals, *6;* migration of, 17; oil spills and, *act.* 137; relocation of, 143

Bison, 9, *9*

Breeding: crossbreeding, 128

Butterflies, 23

C

Cactus, 12, *12*

Captive population, 142, *142*

Carbon cycle, *48,* **49**

Carbon dioxide: carbon cycle and, *48,* 49; greenhouse effect and, 104, *104, act.* 111

Carbon monoxide, 106

Career, 28–29, 87, 119, 147

Carnivores, 21, *21,* 51, *51*

Carrying capacity, 15

Cell(s): solar, 100, *100*

Chemistry Integration, 21, 79

Chemosynthesis, 20, 50–**51,** *50*

Chlorofluorocarbons (CFCs), 105

Chlorophyll, 20

Climate, 41; as abiotic factor in environment, 41–42, *41, 42;* comparing differences in, *act.* 35; global warming and, 136; greenhouse effect and, 104, *104, act.* 111; solar radiation and, 105, *105*

Climax community, 67, *67,* 68

Clown fish, 22, *22*

Coal, 96, *96*

Commensalism, 22, *22*

Communities, 10; climax, 67, *67,* 68; interactions within, 10, 20–24; symbiosis in, **22**–24, *22, 23, 24*

Composting, 115, *115*

Condensation, 45, *45*

Conservation: of fossil fuels, 96; loss of animal habitats and, 133; recycling, 113–115, *113, 115;* reducing use of natural resources, 112–117; reusing, 112, *112*

Conservation biology, 138–143
Consumers, 21, *21,* 51, *51*
Convention on International Trade in Endangered Species, 139
Cooperation, 24
Coral reef, 81, *81*
Corn, 128
Cotton, *94*
Crickets, 12, 13
Crossbreeding, 128
Cycles, 44–49; carbon, *48,* **49;** nitrogen, **46**–47, *46, 47;* water, **44**–**45,** *44, 45*

D

Dam, 97
Decomposers, 21, *21*
Deer, *70*
Desert(s), 8, *8,* **74,** *74;* competition in, 12, *12;* water in, *37*
Desertification, 74
Design Your Own Experiment, 26–27
Diseases: plants as cure for, 128, *128*

E

Eagle, *139*
Earth: biosphere of, 8–9, *8*
Earth Science Integration, 51, 74, 140
Ecological succession, 64–67, *66*
Ecology, 9
Ecosystems, 9, *9,* 62–85; aquatic, 77–83, *act.* 84; carrying capacity of, 15; changes in, 64–67, *64, 65, 66;* competition in, 12, *12;* habitats in, 11, *11,* 12, *12,* 23, *23,* 133–134, *133,* 140–141, *140, 141;* land, 68–76, *68, act.* 76; limiting factors in, 15, *15;* populations in, 10, 13–19, *13, 14, act.* 26–27; stability of, 129, *129*
Electricity: generating, 97–100,
97, 98; water and, 97; wind-mills and, 98, *98*
Elephant, 139
Elevation: and temperature, 40, *40*
Endangered species, 131, *131, 132,* 139, *139*
Endangered Species Act of 1973, 139
Energy: converting, 50–51, *50;* flow of, 50–53; in food chain, 51, *51;* geothermal, **99,** *99;* kinetic, 97; loss of, 53, *53;* nuclear, **98,** *98;* obtaining, 20–21, *20, 21;* photosynthesis and, 20, 50; potential, 97; solar, 99–100, *99, 100, act.* 101, 116–117; transfer of, 51–52, *51, 52*
Energy pyramids, 52–**53,** *53*
Environment, 124–145; abiotic factors in, 36–42, *36;* biodiversity in, 126–136, *126;* biotic factors in, 36; conservation biology and, 138–139, *138;* recognizing differences in, *act.* 125
Erosion, *act.* 93, **109,** *109*
Estuaries, 82–83, *83*
Evaporation, 44, *44, 45*
Everglades, *79*
Explore Activity, 7, 35, 63, 93, 125
Exponential growth, 19, *19*
Extinction: mass, 130, *130*
Extinct species, 130, *130*

F

Field Guide: Biomes Field Guide, 156–159; Waste Management Field Guide, 160–163
Florida Everglades, *79*
Foldables, 7, 35, 63, 93, 125
Food chain(s), 21, *21;* energy in, 51, *51*
Food web, 52, *52*
Forests, *124. see also* Rain
forests; biodiversity in, *126;* as climax community, 67, *67,* 68; as renewable resource, 94, *94;* temperate deciduous, *68,* 70–**71,** *70*
Formaldehyde, 106
Fossil fuels, 96, *96;* alternatives to, 97–99; conservation of, 96; greenhouse effect and, 104
Freshwater ecosystems, 77–79; lakes and ponds, 78–79, *78;* rivers and streams, 77–78, *77;* wetlands, 79, *79, act.* 84–85
Fruit flies: population growth in, *act.* 26–27
Fungi: mutualism and, 22, *22;* niche of, *23*

G

Gas: natural, 96
Geothermal energy, 99, *99*
Glass: recycling, 114
Global warming, 136
Grass: life in, *act.* 7
Grasslands, 75, *75*
Gravity: and atmosphere, 41
Greenhouse effect, 104, *104, act.* 111
Groundwater: pollution of, 108, *108*
Growth: exponential, 19, *19;* of plants, *act.* 54–55; of population, 16–19, *17, 18, 19, act.* 26–27

H

Habitat(s), 11, *11,* 12, *12,* 23, *23;* divided, 134; loss of, 133, *133;* preserving, 140, *140*
Habitat restoration, 141, *141*
Handbooks. *see* Math Skill Handbook, Reference Handbook, Science Skill Handbook, and Technology Skill Handbook
Hazardous wastes, 110, *110*
Health Integration, 106
Herbivores, 21, *21,* 51, *51*
Humus, 38, *act.* 43

Index

Hydroelectric power, **97**
Hydrothermal vents, 51

Iceland: geothermal energy in, 99, *99*
Indoor air pollution, 106, *106*
Insects: competition and, 12; counting population of, 13; interactions of, *6;* niches of, 23, *23,* 24
Internet. *see* Science Online and Use the Internet
Intertidal zone, **82,** *82*
Introduced species, **134,** *134*
Iron: as nonrenewable resource, *95*

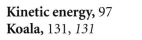

Kinetic energy, 97
Koala, 131, *131*

Labs. *see* Activities, MiniLABs, and Try at Home MiniLABs
Lakes, 78–79, *78*
Land biomes, 68–76, *68, act.* 76; deserts, 74, *74;* grasslands, 75, *75;* taiga, 70, *70;* temperate deciduous forests, 68, 70–71, *70;* temperate rain forests, 71, *71;* tropical rain forests, 68, 72–73, *72, 73;* tundra, 69, *69*
Landfills: 110, *110*
Latitude: and temperature, 39, *39*
Lava, *62*
Law(s): on endangered species, 139
Lichens: and mutualism, 22, *22*
Life: variety of. *see* Biodiversity
Light: as abiotic factor in environment, 38, *38*
Limiting factors, 14, *15*
Lynx, *70*

Maize, 128
Malaria, *128*
Manatee, 138, *138*
Mars, 9
Math Skill Handbook, 182–188
Math Skills Activities, 40, 80, 129
Measurement: of biodiversity, 126–127, *127*
Medicine: plants as, 128, *128*
Mercury, 9
Metal(s): as nonrenewable resource, 95, *95;* recycling, 114
Migration, 17
Millipedes, *23*
Mineral(s): as nonrenewable resource, 95, 96
MiniLABs: Comparing Biotic Potential, 17; Comparing Fertilizers, 47; Modeling Freshwater Environments, 78; Measuring Acid Rain, 103; Testing the Effects of Acid Rain, 135
Mining, 96
Model and Invent, 116–117
Mountains: rain shadow effect in, 42, *42;* temperature and, 40, *40*
Movement: of populations, 17, *17*
Mutualism, **22,** *22*

National Geographic Visualizing: Population Growth, 18; The Carbon Cycle, 48; Secondary Succession, 66; Solar Energy, 101; Threatened and Endangered Species, 132
Native species, **134,** *134*
Natural gas, 96
Natural resources, **94**–100. *see also* Resources
Niche, **23**–24, *23*
Nitrogen cycle, **46**–47, *46, 47*
Nitrogen dioxide, 103

Nitrogen fixation, **46,** *46*
Nonrenewable resources, **95,** *95*
Nuclear energy, **98,** *98*
Nuclear waste, 98, 110

Ocean water, 108, *108*
Oil: and pollution, 108, *108, act.* 137. *see also* Petroleum
Omnivores, 21, *21,* 51, *51*
Oryx, 142, *142*
Oxygen: and respiration, 37
Ozone depletion, **105,** *105,* 136
Ozone layer, **136**

Paper: recycling, 115
Parasitism, **22,** *22*
Pelican, 143, *143*
Permafrost, 69, *69*
Pesticides, 143
Petroleum, **95,** 96
Photosynthesis, 20, 22, 37, 38, *38,* 50
Physics Integration, 41, 97
Pigeon, 130, *130*
Pioneer species, **64,** *65*
Plains, 75
Planaria: feeding habits of, *act.* 25
Plant(s): diversity of, 128, *128, act.* 144–145; growth of, *act.* 54–55; houseplants, *act.* 23; movement of, 17, *17;* nitrogen fixation and, 46, *46;* photosynthesis in, 20, 37, 38, *38,* 50; seed banks for, 143
Plastics: recycling, 113, *113*
Polar bears, 8, *8*
Poles: of Earth, 39, *39*
Pollutants, **102**
Pollution, 102–110, 135–136; of air, 102–106, *102, 103, 104, 105, act.* 111, 135, *135;* nuclear power and, 98; of soil, 109–110, *110;* of water, 107–108, *107, 108,* 135, 143, *act.* 137

Ponds, 78–79, *78*
Population(s), **10;** biotic potential of, 16, 17; captive, **142,** *142;* growth of, 16–19, *17, 18, 19, act.* 26–27; movement of, 17, *17;* size of, 13–16, *13, 14, 15*
Potential energy, 97
Power: hydroelectric, **97;** nuclear, 98, *98;* wind, 98
Prairies, 75
Precipitation: acid, **103,** *103,* **135,** *135*
Predators, 24, *24*
Prey, 24, *24*
Primary succession, 64–65, *65,* 67
Problem-Solving Activities, 15, 114
Producers, 20, *20,* 51, *51*
Ptarmigan, *69*

R

Rabbits, 13, 14
Radiation: from the Sun, 105, *105*
Radioactive waste, 98, 110
Radon, 106, *106*
Rain: acid, 103, *103,* **135,** *135;* water pollution and, 107, *107*
Rain forests: life in, 8; temperate, **71,** *71;* tropical, 68, **72–73,** *72, 73;* water in, *37*
Rain shadow effect, 42, *42*
Recycling, 113–115, *113, 115*
Reducing, 112
Reef, 81, *81*
Reference Handbook, 189–194
Reintroduction programs, **142**
Renewable resources, **94**–95, *94, 95*
Resources: conservation of, 112–117; natural, **94**–100; nonrenewable, **95,** *95;* renewable, **94**–95, *94, 95*
Respiration: and oxygen, 37
Reusing, 112, *112*
Rhinoceros, 131, *131*

River(s), 77
Roundworm: as parasite, 22, *22*

S

Saltwater ecosystems, 80–83; coral reefs, 81, *81;* estuaries, 82–83, *83;* oceans, 81; seashores, 82, *82*
Savannas, 75, *75*
Science and History. *see* TIME
Science and Language Arts, 118–119
Science and Society. *see* TIME
Science Online: Data Update, 16, 41, 65, 139; Research, 10, 49, 81, 104, 114, 142
Science Skill Handbook, 164–177
Science Stats, 56–57
Scorpion, *74*
Sea anemone, 22, *22*
Seashores, 82, *82*
Secondary succession, 65, 67
Seed(s): movement of, 17, *17*
Seed banks, 143
Sheep, 17
Smog, 102, *102*
Smoking: and indoor air pollution, 106
Soil, **38;** as abiotic factor in environment, 38, *act.* 43; building, 64–65, *65;* determining makeup of, 38; loss of, *act.* 93, 109, *109;* nitrogen in, 47, *47;* pollution of, 109–110, *110;* topsoil, *act.* 93, 109, *109;* in tropical rain forests, 72–73
Solar cells, 100, *100*
Solar cooking, *act.* 116–117
Solar energy, 99–100, *99, 100, act.* 101, 116–117
Solid waste, 109
Species: endangered, **131,** *131, 132,* 139, *139;* extinct, **130,** *130;* introduced, **134,** *134;* native, **134,** *134;* pioneer, **64,** *65;* threatened, **131,** *131, 132*
Spiders, *23*

Standardized Test Practices, 33, 61, 91, 123, 151
Steel: recycling, 114
Stream(s), 77–78, *77*
Succession, 64–67, *66;* primary, 64–65, *65,* 67; secondary, 65, 67
Sulfur, 103
Sulfur dioxide, 103
Sun: radiation from, 105, *105*
Surface water, 107, *107*
Symbiosis, **22**–24, *22, 23, 24*

T

Taiga, **70,** *70*
Taxol, *128*
Technology: turbine, 97, *97,* 98, *98*
Technology Skill Handbook, 178–181
Temperate deciduous forests, 68, **70–71,** *70*
Temperate rain forests, **71,** *71*
Temperature: as abiotic factor in environment, 39–40, *39, 40;* of air, 104, *104;* of oceans, 80
Termites, *23*
Test Practices. *see* Standardized Test Practices
The Princeton Review. *see* Standardized Test Practice
Threatened species, **131,** *131, 132*
TIME: Science and History, 28–29; Science and Society, 86–87, 146–147
Topsoil: loss of, *act.* 93, 109, *109*
Traditional Activity, 54–55
Transpiration, 44
Tropical rain forests, 68, **72–73,** *72, 73. see* Rain forests
Try at Home MiniLABs: Observing Seedling Competition, 13; Determining Soil Makeup, 38; Modeling Rain Forest Leaves, 72; Observing Mineral Mining Effects, 96; Demonstrating Divided Habitats, 133

Index

Tundra, 69, *69*
Turbine, 97, *97,* 98, *98*

U

Use the Internet, 84–85

V

Venus, 9
Visualizing. *see* National
    Geographic

W

Wastes: hazardous, **110,** *110;*
    radioactive, 98, 110; solid, 109
Water: as abiotic factor in
    environment, 37, *37;* in
    generation of electricity, 97;
    groundwater, 108, *108;* from
    hydrothermal vents, 51; as
    limiting factor in ecosystem,
    15; in oceans, 108, *108;*
    pollution of, 107–108, *107,*
    108; surface, 107, *107*
Water cycle, 44–45, *44, 45*
Water pollution, 135,
    act. 137, 143
Wetlands, 79, *79, act.* 84–85
Wildebeests, *14*
Wildlife corridors, 140, *140*
Wildlife management, 141
Wind, 41, *41*
Windmills, 98, *98*
Woodpeckers, 11, *11,* 12, *12*

Y

Yellowstone National Park, 140

Index

Bauer/TSA; **70** (tl)Peter Ziminski/VU, (tr)Leonard Rue III/VU, (bl)C.C. Lockwood/DRK, (br)Larry Ulrich/DRK; **71** (t)Fritz Polking/VU, (b)William Grenfell/VU; **72** Lynn M. Stone/DRK; **74** (l)Joe McDonald/DRK, (r)Steve Solum/AMP; **75** Kevin Schafer; **77** W. Banaszewski/VU; **78** (l)Dwight Kuhn, (r)Mark E. Gibson/VU; **79** James R. Fisher/DRK; **80** D. Foster/WHOI/VU; **81** (l)C.C. Lockwood/AMP, (r)Steve Wolper/DRK; **82** (t)Dwight Kuhn, (c)Glenn Oliver/VU, (b)Stephen J. Krasemann/DRK; **83** (l)John Kaprielian/PR, (r)Jerry Sarapochiello/AMP; **84** (t)Dwight Kuhn, (b)John Gerlach/DRK; **85** Fritz Polking/AMP; **86 87** Courtesy Albuquerque Public Schools; **88** (tl)Bob Gurr/DRK, (tr)M.C. Chamberlain/DRK, (b)David Pearson/VU; **89** (l)James P. Rowan/DRK, (r)John Shaw/TSA; **92** Rannels from GH; **92-93** GH; **93** AMP; **94** (l)Keith Lanpher/LA, (r)Richard Thatcher/David R. Frazier Photolibrary; **95** (t)Solar Cookers International, (bl)Brian F. Peterson/TSM, (br)Ron Kimball Photography; **96** Larry Mayer/LA; **99** (t)Torleif Svenson/TSM, (c)Les Gibbon/Cordaiy Photo Library Ltd./CB, (b)Rob Williamson; **100** Sean Justice; **101** (t)Lowell Georgia/Science Source/PR, (cl)NASA, (c)CB, (cr)Sean Sprague/Impact Visuals/PQ, (bl)Lee Foster/AMP, (br)Robert Perron; **102** Philippe Renault/LA; **103** (l)NYC Parks Photo Archive/FP, (r)Kristen Brochmann/FP; **106** AH; **107** (l)Jeremy Walker/Science Photo Library/PR, (c)John Colwell from GH, (r)Telegraph Colour Library/FPG; **109** (tl)Larry Mayer/LA, (tr)ChromoSohm/TSM, (c)David R. Frazier Photolibrary, (b)Inga Spence/VU; **110** (tl)AMP, (tr)Andrew Holbrooke/TSM, (bl)IC, (br)AH; **112** Paul A. Souders/CB; **113** IC; **116** Larry Lefever from GH; **116** (t)Howard Buffett from GH, (b)Solar Cookers International; **117** John D. Cunningham/VU; **118** Frank Cezus/FPG; **118** Wilford Haven/LA; **119** Nic Paget-Clarke; **120** (t)MB, (c)David R. Frazier Photolibrary, (b)George Diebold/TSM; **121** (l)Lebrum/LA, (r)Spencer Grant/PE; **122** (l)James N. Westwater, (r)Steve McCutcheon/VU; **124** Gordon Wiltsie/Peter Arnold, Inc.; **124-125** Bob Firth; **125** Geoff Butler; **126** (l)Andy Sacks/Stone, (r)Erwin & Peggy Bauer/AMP; **127** (t)Leonard Lee Rue III/PR, (c)Chuck Pefley/Stone, (b)Charles W. Mann/PR; **128** (l)Gilbert S. Grant/PR, (lc)Rexford Lord/PR, (rc)Ray Pfortner/Peter Arnold, Inc., (r)William D. Adams; **129** William D. Adams; **131** (t)Joe McDonald/TSA, (b)Len Rue, Jr./AA; **132** (bkgd) Kennan Word/CB, (tl)Joel Sartore, (tr)William H. Amos, (c)Joseph Van Wormer/AMP/PQ, (b)Erwin & Peggy Bauer/AMP/PQ; **133** (l)Jim Baron/The Image Finders, (r)Melissa Hathaway/ODNR Div. of Wildlife; **134** (l)University of MN Sea Grant Program, (r)Marcia Griffen/ES; **135** SuperStock; **137** A. Greth/BIOS/Peter Arnold, Inc.; **138** Michael Pogany/Columbus Zoo; **139** (l)Bill Pogue/Stone, (r)Daniel J. Cox/Stone; **141** Save The Bay® People for Narragansett Bay; **142** Leonard L.T. Rhodes/AA; **143** Hal Beral/VU; **144** Zig Leszczynski/ES; **146** (t)Jacques Jangoux/PR, (b)Michael Fogden/DRK; **147** F. Gilson/Bios/Peter Arnold, Inc.; **148** (t)David Woodfall/Stone, (c)David Campbell/VU, (b)Kenneth W. Fink/AMP; **149** (l)Inga Spence/VU, (c)Bogart Photography, (r)Juan Manuel Renjifo/ES; **154-155** PD; **156** Greg Probst/Stone; **157** (t)GH, (b)George Ranalli/PR; **158** (t)AH, (b)Tom Bean/Stone; **159** (t)Tom Bean/DRK, (b)Gary Braasch/Stone; **160** Tony Freeman/PE/PQ; **161** (t)Dominic Oldershaw, (b)Mary Kate Denny/PE; **162** (t)file photo, (b)John Evans; **163** (t)KS, (b)Ken Lax; **164** Timothy Fuller; **168** First Image; **171** Dominic Oldershaw; **172** StudiOhio; **173** First Image; **175** Richard Day/AA; **178** Paul Barton/TSM; **181** Charles Gupton/TSM; **190** MM; **191** (t)NIBSC/Science Photo Library/PR, (bl)Dr. Richard Kessel, (br)David John/VU; **192** (t)Runk/Schoenberger from GH, (bl)Andrew Syred/Science Photo Library/PR, (br)Rich Brommer; **193** (t)G.R. Roberts, (bl)Ralph Reinhold/ES, (br)Scott Johnson/AA; **194** Martin Harvey/DRK.

Acknowledgements

"Beauty Plagiarized" by Amitabha Mukerjee. Reprinted by permission of the author.

PERIODIC TABLE OF THE ELEMENTS

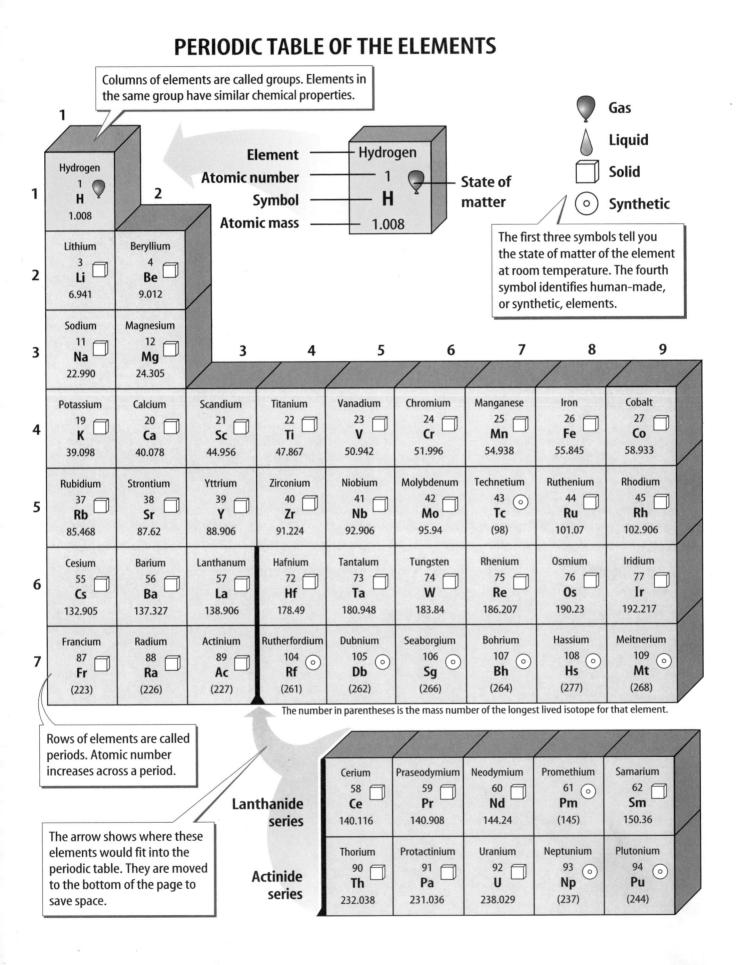

Columns of elements are called groups. Elements in the same group have similar chemical properties.

Element — Hydrogen
Atomic number — 1
Symbol — H
Atomic mass — 1.008

State of matter

Gas
Liquid
Solid
Synthetic

The first three symbols tell you the state of matter of the element at room temperature. The fourth symbol identifies human-made, or synthetic, elements.

1	2	3	4	5	6	7	8	9
1 Hydrogen 1 **H** 1.008								
2 Lithium 3 **Li** 6.941	Beryllium 4 **Be** 9.012							
3 Sodium 11 **Na** 22.990	Magnesium 12 **Mg** 24.305							
4 Potassium 19 **K** 39.098	Calcium 20 **Ca** 40.078	Scandium 21 **Sc** 44.956	Titanium 22 **Ti** 47.867	Vanadium 23 **V** 50.942	Chromium 24 **Cr** 51.996	Manganese 25 **Mn** 54.938	Iron 26 **Fe** 55.845	Cobalt 27 **Co** 58.933
5 Rubidium 37 **Rb** 85.468	Strontium 38 **Sr** 87.62	Yttrium 39 **Y** 88.906	Zirconium 40 **Zr** 91.224	Niobium 41 **Nb** 92.906	Molybdenum 42 **Mo** 95.94	Technetium 43 **Tc** (98)	Ruthenium 44 **Ru** 101.07	Rhodium 45 **Rh** 102.906
6 Cesium 55 **Cs** 132.905	Barium 56 **Ba** 137.327	Lanthanum 57 **La** 138.906	Hafnium 72 **Hf** 178.49	Tantalum 73 **Ta** 180.948	Tungsten 74 **W** 183.84	Rhenium 75 **Re** 186.207	Osmium 76 **Os** 190.23	Iridium 77 **Ir** 192.217
7 Francium 87 **Fr** (223)	Radium 88 **Ra** (226)	Actinium 89 **Ac** (227)	Rutherfordium 104 **Rf** (261)	Dubnium 105 **Db** (262)	Seaborgium 106 **Sg** (266)	Bohrium 107 **Bh** (264)	Hassium 108 **Hs** (277)	Meitnerium 109 **Mt** (268)

The number in parentheses is the mass number of the longest lived isotope for that element.

Rows of elements are called periods. Atomic number increases across a period.

The arrow shows where these elements would fit into the periodic table. They are moved to the bottom of the page to save space.

Lanthanide series	Cerium 58 **Ce** 140.116	Praseodymium 59 **Pr** 140.908	Neodymium 60 **Nd** 144.24	Promethium 61 **Pm** (145)	Samarium 62 **Sm** 150.36
Actinide series	Thorium 90 **Th** 232.038	Protactinium 91 **Pa** 231.036	Uranium 92 **U** 238.029	Neptunium 93 **Np** (237)	Plutonium 94 **Pu** (244)